Contents

TARGETING SCIENCE YEAR 5 © PASCAL PRESS ISBN: 9781925726541

Targeting Science Year 5

ISBN: 9781925726541

Published by Pascal Press
PO Box 250
Glebe NSW 2037
www.pascalpress.com.au
contact@pascalpress.com.au

This edition of *Skill Sharpeners: Science* is published by arrangement with Evan-Moor Corporation, USA.

For sale in Australia and New Zealand.

Authors: Guadalupe Lopez, Lisa Vitarisi Mathews
Publisher: Lynn Dickinson
Cover design: Janice Bowles
Illustrator: Paul Lennon, *www.dreamstime.com.au*
Editor Australian edition: Stella Tarakson
Typesetter: Stacey Grainger

Printed in China by 1010 International Ltd.

Introduction

Welcome to your Year 5 *Targeting Science* activity book! It is packed with interesting and exciting activities to help you understand and enjoy science at home or school.

Targeting Science has been written to support the Australian Primary Science Curriculum Version 9.0 and is divided between:

- Biological Sciences
- Earth & Space Sciences
- Physical Sciences
- Chemical Sciences

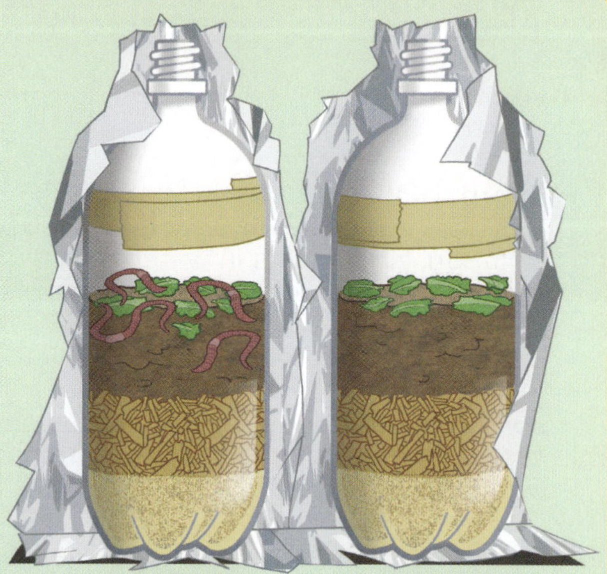

The Science Understanding and Inquiry Skills elements (including Science as a Human Endeavour) are embedded.

Hands-on activities provide opportunities to bring the science concepts to life. The exercises develop process skills such as observing, collecting, recording and organising information and making models of scientific events that happen in the natural world. All activities use easily sourced, inexpensive items.

Topics are introduced with explanations plus images to provide background information. Some lessons include a QR code where you can access a video for further explanations. You will be challenged to match, sort, label, sequence, analyse and answer questions with regular vocabulary practise puzzles throughout the book to help familiarise you with scientific terms. See the Glossary on the next page for some terms you may not already know. Answers are included at the end of the book.

FREE Teaching Guide.

This QR code links to a downloadable PDF of a Teaching Guide to support the material in this student workbook. The guide contains:

- Teaching plans and checklists.
- Graphic organisers.
- Material request forms (to send home for parents).
- Background information about each major topic as well as extension activities.

Use this QR code to access the FREE Teaching Guide.

Front cover – have you looked at the front cover? It depicts what life was like before a significant scientific discovery that led to the development of new technology. What is the difference between Science and Technology? The following quote sums it up nicely.

Science is the process of acquiring knowledge of natural phenomenon along with various reasons. Technology is the application of Science to the solution of problems. (Sismondo, 2018)

Glossary

- **biomimicry** - the practice of making technological and industrial design copy natural processes.
- **compression** - the act of pressing something into a smaller space or putting pressure on it from different sides until it gets smaller.
- **interparticle / intermolecular space** - The space between two molecules in any substance is called intermolecular space or interparticle space. When we make a solution, e.g. alcohol in water, a part of the solute goes in the interparticle space. Intramolecular space is the space present within a molecule.
- **opaque** – a material that does not let light through.
- **soil erosion** - The process by which soil, rocks and other surface materials of Earth are worn away, such as by the action of water, glaciers, wind or waves.
- **translucent** – a material that transmits some light but is not transparent, such as frosted glass.
- **transparent** – a material that lets light through, such as glass.

SAFETY
All of the investigations in this book are designed for kids to do safely in the home or classroom. However, adult supervision is recommended.

TARGETING SCIENCE YEAR 5 © PASCAL PRESS ISBN: 9781925726541

Year 5 Australian Science Curriculum Correlations

ACARA Code	Content description	Strand	Sub strand	Pages
AC9S5U01	Students learn to examine how particular structural features and behaviours of living things enable their survival in specific habitats	Science Understanding	Biological Sciences	2-31
AC9S5U02	Students learn to describe how weathering, erosion, transportation and deposition cause slow or rapid change to Earth's surface	Science Understanding	Earth & Space Sciences	32-58
AC9S5U03	Students learn to identify sources of light, recognise that light travels in a straight path and describe how shadows are formed and light can be reflected and refracted	Science Understanding	Physical Sciences	59-81
AC9S5U04	Students learn to explain observable properties of solids, liquids and gases by modelling the motion and arrangement of particles	Science Understanding	Chemical Sciences	82-96
AC9S5H01	Students learn to examine why advances in science are often the result of collaboration or build on the work of others	Science as human endeavour	Nature and development of science	6, 11, 12, 19, 20, 24, 41, 47, 54, 60, 82, 95
AC9S5H02	Students learn to investigate how scientific knowledge is used by individuals and communities to identify problems, consider responses and make decisions	Science as human endeavour	Use and influence of science	3, 12, 36, 43, 45, 47, 55, 83
AC9S5I01	Students learn to pose investigable questions to identify patterns and test relationships and make reasoned predictions	Science Inquiry	Questioning and Predicting	8, 16, 39, 49, 57, 80
AC9S5I02	Students learn to plan and conduct repeatable investigations to answer questions, including, as appropriate, deciding the variables to be changed, measured and controlled in fair tests; describing potential risks; planning for the safe use of equipment and materials; and identifying required permissions to conduct investigations on Country/Place	Science Inquiry	Planning and conducting	8, 16, 39, 49, 62, 66, 68, 69, 71, 79, 86, 93
AC9S5I03	Students learn to use equipment to observe, measure and record data with reasonable precision, using digital tools as appropriate	Science Inquiry	Planning and conducting	24, 40, 71, 72
AC9S5I04	Students learn to construct and use appropriate representations, including tables, graphs and visual or physical models, to organise and process data and information and describe patterns, trends and relationships	Science Inquiry	Processing, modelling and analysing	6, 10, 17, 34, 39, 40, 50, 72, 84, 89, 94
AC9S5I05	Students learn to compare methods and findings with those of others, recognise possible sources of error, pose questions for further investigation and select evidence to draw reasoned conclusions	Science Inquiry	Evaluating	26, 50, 72, 86, 88, 94
AC9S5I06	Students learn to write and create texts to communicate ideas and findings for specific purposes and audiences, including selection of language features, using digital tools as appropriate	Science Inquiry	Communicating	17, 18, 27, 73, 87, 90, 91

Dolphin Adaptations

Concept:

External characteristics of living things allow their needs to be met.

Define It!

adaptation: a feature that helps a living thing stay alive

blowhole: an opening on the top of the head used for breathing

survive: to stay alive

trait: a feature belonging to a living thing

Living things have traits, or features. A trait that helps a living thing survive is called an adaptation.

Physical, or structural adaptations are the things that animals **have** that help them to survive, for example dolphins have a special adaptation that helps them breathe under water. A dolphin breathes through a blowhole on the top of its head, instead of through a nose or mouth like ours.

The dolphin uses strong muscles to open its blowhole when it swims to the water's surface. As it dives underwater, the blowhole closes.

Dolphins have other structural adaptations too. They have a strong flexible tail, smooth skin and two flippers. They glide through the water by moving their tail and steering with their flippers.

Behavioural adaptations are the things animals **do** to help them survive. Dolphins often choose to live in groups (called pods) to help them stay safe and catch food.

Label these dolphin adaptations as structural or behavioural:

1. Blowhole for breathing when a dolphin comes to the surface: ______________________

2. Smooth skin to glide through water: ______________________

3. Works with other dolphins to round up shoals of fish: ______________________

TARGETING SCIENCE YEAR 5 © PASCAL PRESS ISBN: 9781925726541

Rock Pocket Mouse

https://clickv.ie/w/Exgx

Use this QR code to access a video on this topic.

Define It!

camouflage: colours that hide an animal

lava: melted rock

volcano: an opening in the earth where lava flows out

Colour is an adaptation that helps animals survive. When animals hide by blending in with the things around them, it is called **camouflage**. The rock pocket mouse is an example of camouflage. This mouse lived in the desert. Its sandy-brown colour blended in with the sand and rocks and camouflaged the mouse. Owls and other animals that eat mice couldn't easily see it, so the rock pocket mice had a good chance of surviving. Then, long ago, **lava** from a **volcano** flowed over part of the desert. The lava cooled into dark rock. The light-coloured mice were no longer safe on the dark rocks. By chance, a few dark-coloured mice were born. The dark-coloured mice survived, and the light-coloured mice were eaten. More dark mice were born. Now, almost all of the rock pocket mice living on the dark rocks are dark-coloured, and the mice living in the sandy part of the desert are light-coloured.

photo by Roger W. Barbour

photo by R. B. Forbes, © American Society of Mammalogists

Answer the questions.

1. How does the rock pocket mouse hide? ______________________

____

2. Which mice survive on the dark rocks?

____

Concepts:

For any particular environment, some kinds of living things survive well, some survive less well, and some cannot survive at all.

Living things have characteristics that allow them to survive.

Cactus Spines

Concepts:

Living things have characteristics that allow them to survive.

Living things have different characteristics that allow their needs to be met.

Define It!

protect: to keep safe

reproduce: to produce offspring

spines: the stiff, sharp needles of a cactus

A cactus has adaptations that help it survive in a hot, dry desert. One adaptation is its needles, or **spines**. A spine is an adaptation of a leaf. Spines can be useful in different ways. Spines catch water and drip it toward the roots of the cactus. A cactus may be thickly covered with spines that shade it and keep it from drying out in the sun. Some spines **protect** a cactus from animals looking for a juicy meal. The animals learn that the spines will stick in their mouth, so they keep away. The jumping cholla (CHOY-ah) cactus uses its spines to **reproduce**. If an animal brushes against a jumping cholla, a small piece breaks off and sticks to the animal. In time, the piece drops to the ground, grows roots, and becomes a new plant.

Circle *true* or *false*.

1. A cactus spine adapted from a leaf. **true** **false**
2. Spines are not very useful to a cactus. **true** **false**
3. One type of cactus uses its spines to make new plants. **true** **false**

TARGETING SCIENCE YEAR 5 © PASCAL PRESS ISBN: 9781925726541

Barrel Cactus

Cactus spines catch water to help a cactus survive. The barrel cactus has other adaptations that help it get and store water. The roots of the cactus spread out like a net to quickly drink up even the smallest amount of desert rain. A barrel cactus has folds that can swell and store litres of water. Its thick, waxy skin keeps the water inside the plant from escaping into the air.

Label the diagram to show four adaptations of the barrel cactus. Use the words below.

roots   folds   spines   thick, waxy skin

1. When rain falls in the desert, it dries up quickly. That is why the roots ____________________

2. A cactus may have to live without rain for a long time. That is why the folds ____________________

Skills:

Identify the external characteristics of different kinds of plants and animals that allow their needs to be met.

Analyse and interpret information presented in a visual format.

Adaptations

Desert Adaptations

https://clickv.ie/w/0wgx

Use this QR code to access a video on this topic.

For many First Nations people whose Country is situated on dry arid or desert environments, knowing how and where to find water is very important. The water-holding frog is found in creeks, swamps and claypans in these regions. The frog has developed the adaptation to be able to store water underneath its skin, which it can then re-absorb when there has been no rain.

The water-holding frog absorbs water in the wet season, then burrows underground to find cooler temperatures. Once underground, it sheds several layers of skin to create a translucent, waterproof cocoon around itself. This helps it preserve water for the many long months ahead.

First Nations people found these frogs underground by looking for markings on the surface of the ground or by tapping the ground with the end of a spear.

When the frogs were found, they were dug up, and they were squeezed gently to release water from underneath the skin. The water was clean enough for humans to be able to drink it.

First Nations people in desert regions understand the water holding adaptation of this species and in times of drought or emergency these adaptations can be used as a vital water source. This knowledge is still taught by the Traditional Owners in these regions to ensure that this lifesaving skill is passed on.

Sort these adaptations of water holding frogs into structural and behavioural adaptations.

Frog digs under the ground after rain.	Frog skin is able to hold water.	Frog has strong back legs to help it burrow underground.
Frog only emerges when it has rained.	Male water holding frogs make loud, long, slow 'maaaw-w-w' sounds to attract females.	Frog has strong webbing between toes for more speed when hunting under water.

Structural Adaptations: (things the frog has got that help it to survive)	**Behavioural Adaptations:** (things the animal does to help it to survive)

Adaptations

TARGETING SCIENCE YEAR 5 © PASCAL PRESS ISBN: 9781925726541

Review the Words

Skill: Apply content vocabulary in context sentences.

Adaptations

Read each clue and write the missing word on the lines.

camouflage	volcano	adaptation
blowhole	spines	survive

1. A dolphin breathes through its ______________________.

2. The dark colour of a rock pocket mouse is an ______________.

3. The sharp needles of a cactus are called ______________.

4. Adaptations help living things ______________________.

5. Animals use ______________________ to help them hide.

6. Lava flows from a ______________________.

What are some adaptations a kangaroo might have to survive in Australia?

____

____

____

____

Plant Adaptations

Skills:

Gather and interpret scientific data from observation.

Use observations to construct an explanation.

A cactus can store water for months. But not all plants have this adaptation. See what happens to a celery stalk with and without water.

What You Need

- celery stalk
- paper towel
- water in a glass
- an adult to cut the celery

What You Do

1. Have an adult cut off the bottom of the celery stalk. Notice the tiny holes at the bottom. These are tiny tubes that go up the stalk.
2. Set the stalk on a paper towel and leave it out overnight.
3. Look at the celery stalk. Notice how it looks and feels now.
4. Stand the celery in the glass of water, with the cut end at the bottom. Leave it like this overnight.
5. Now look at the celery stalk. Notice how it looks and feels.

What Did You Discover?

1. Describe how the celery looked and felt after being on the paper towel.

___

2. Describe how the celery looked and felt after being in water overnight.

___

3. Compare the celery with a cactus. Write one way they are alike and one way they are different.

___

___

TARGETING SCIENCE YEAR 5 © PASCAL PRESS ISBN: 9781925726541

What's My Adaptation?

Pretend that you are a dolphin, a rock pocket mouse, or a cactus. Explain what your adaptations are and how they help you survive. Then draw and label a diagram that shows your adaptations.

Hint

Dolphins have adapted to living underwater. The rock pocket mouse has adapted to the colour of the dark rocks. Cactus plants have adapted to living in a hot, dry desert.

Draw

Unique Environments

Concepts:

Organisms develop adaptations to fit their unique environments.

Define It!

adaptations: changes in an organism that make it better able to live in a particular place or situation

diet: the kind of food that an organism regularly eats

In order for an ecosystem to work, organisms that are part of the community must have their basic needs met. That includes being able to find the right foods for their particular **diets**. Imagine a polar bear trying to live in a bamboo forest or a panda making its home on the Arctic ice.

Although polar bears and pandas are both members of the bear family, they live in very different ecosystems. Giant pandas live in the rainy, mountainous forests of central China. They have a diet almost entirely of bamboo. Polar bears live on the barren ice of the Arctic Circle and hunt seals. These animals would not survive if they swapped places because they have developed **adaptations** that fit their unique environments.

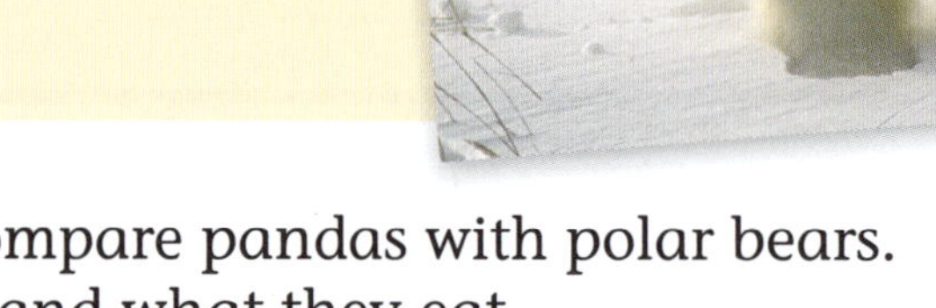

Fill in the diagram to compare pandas with polar bears. Include where they live and what they eat.

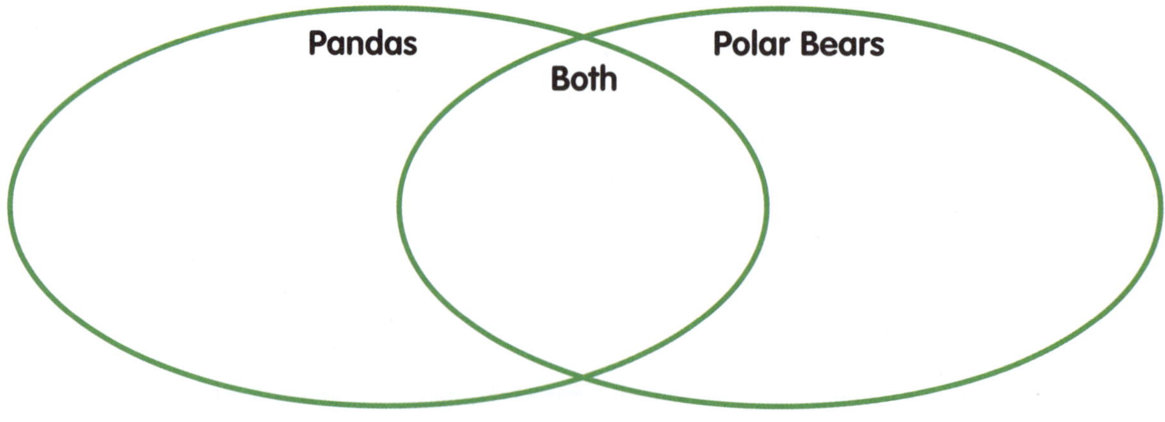

Adaptations

TARGETING SCIENCE YEAR 5 © PASCAL PRESS ISBN: 9781925726541

Define It!

blubber: the fat of sea mammals

habitat: the natural home of an animal, plant, or other organism

specialised: developed for a specific purpose

Some organisms, such as humans, have adaptations that allow them to live through changing conditions. These adaptations help them move between different **habitats** and survive on a wide variety of foods. However, other organisms, such as the panda and the polar bear, have **specialised** adaptations. These adaptations help the animals thrive in their specific environments. However, they would not be useful in a different habitat.

Pandas have developed specialised adaptations for eating the bamboo that grows in their forests. They have flat teeth that help them grind the plants. They also have a bone that extends from their wrist, which helps them hold bamboo shoots and leaves.

Polar bears have developed specialised adaptations for living on the ice and hunting seals. To keep from slipping, the bears have small bumps on the bottom of each paw. Their thick claws grip the ice and help catch their prey. In addition, polar bears' toes are slightly webbed, which helps them swim in the waters between pieces of ice. They also have a 10 cm layer of **blubber**, which keeps them warm in the cold.

Concepts:

Pandas have developed specialised adaptations to help them eat bamboo.

Polar bears have developed specialised adaptations to help them live on the ice and hunt seals.

Answer the questions.

1. Name two specialised adaptations pandas have.

___

___

2. Name three specialised adaptations polar bears have.

___

___

___

TARGETING SCIENCE YEAR 5 © PASCAL PRESS ISBN: 9781925726541

Extinction

Concepts:

If organisms do not adapt to their environment, they are in danger of extinction.

Habitat destruction can lead to extinction, which could impact the ecosystem.

Define It!

extinction: dying out

impact: a strong effect

Polar bears and pandas have very specific needs for survival. Polar bears depend on sea ice in order to hunt and breed. Pandas depend on a single plant for almost all of their food. However, these animals are now being threatened with **extinction**. Pandas are in danger because large parts of their habitats are being destroyed. This means there is less bamboo for them to eat. Polar bears are at risk because the polar ice caps are melting. This means there is less space for them to hunt seals. If these changes continue, it is likely that neither polar bears nor pandas will survive.

What would be the **impact** of this loss on the bears' ecosystems? Scientists are studying this to find out more. But the good news is that the bears' shrinking populations have increased people's awareness of the causes of habitat loss. And this has helped us understand the relationship between all living and nonliving things within any ecosystem—including our own.

Melting ice caps mean polar bears must swim greater distances to reach their hunting grounds.

Why are pandas and polar bears being threatened with extinction?

____

____

____

____

____

 TARGETING SCIENCE YEAR 5 © PASCAL PRESS ISBN: 9781925726541

The Spiny-Leaf Stick Insect belongs to a well camouflaged group of stick insects called Phasmids.

These amazing animals look more like dead leaves than sticks.

The Spiny-Leaf Stick Insect blends into trees and bushes so completely, they are very hard to see in the wild. They hang motionless from branches and feed on fresh leaves, just like other phasmid species.

Spiny-Leaf Stick Insects are found in the north east of Australia, where they feed on a variety of plant types in the wild. They have a very interesting reproductive cycle. The eggs are tossed down from the trees by females to the forest floor. They are small and brown and camouflaged to look very much like plant seeds. For this reason the eggs are often collected by ants and stored below ground in their nests, which also protects the egg from predators.

After hatching, each baby Spiny Leaf Insect (nymph) makes its way to the surface and into a tree. The nymph looks and behaves very much like an ant at this stage of its life cycle. This doesn't fool ants, but does fool other animals such as birds, which don't like eating ants. Because of this clever camouflage, the baby Spiny-Leaf Stick Insects are able to run around and climb up trees with less risk of being eaten by a bird.

1. What are two things that help a Spiny-Leaf Stick Insect blend in so well?

Extinction Link

Skill:

Interpret information in graphic representations.

A word cloud shows information based on the size of different words. In this word cloud, the bigger the word, the closer the animal is to extinction. Use the word cloud to answer the questions below.

numbat

koala King Island scrubtit

Central rockrat thylacine

greater bilby Gilbert's potoroo

human Orange bellied parrot

1. The thylacine is extinct. Which animal is most likely to become extinct next? ____________________
2. If the King Island scrubtit population grows, would you expect the word to become bigger or smaller? ____________________
3. Which two organisms in this word cloud are least likely to become extinct? ____________________ ____________________
4. If the woodlands continue to shrink, would you expect the word koala to become bigger or smaller? ____________________
5. If we added the word *dinosaur* to this word cloud, which word do you think it would be the same size as? ____________________

TARGETING SCIENCE YEAR 5 © PASCAL PRESS ISBN: 9781925726541

Skill:

Apply content vocabulary.

Select from the list of vocabulary words to complete the sentences. Then unscramble the shaded letters to decode the secret message.

specialised	blubber	impact	diet
extinction	adaptations	habitat	

1. A polar bear's __ __ __ __ __ __ __ keeps it warm in the freezing Arctic temperatures.

2. Because of __ __ __ __ __ __ __ loss, both the polar bear and the panda are now threatened with __ __ __ __ __ __ __ __ __ __.

3. Some animals have developed __ __ __ __ __ __ __ __ __ __ __ that are __ __ __ __ __ __ __ __ __ __ __ for their unique environments.

4. A panda's __ __ __ __ is made up almost entirely of bamboo.

Although polar bears spend most of their time on the sea ice of the Arctic Ocean, they also live on the __ __ __ __ __ __, a vast plain of permanently frozen ground.

Adaptations

Blubber Glove

Skills:

Conduct experiments, record data, and analyse results.

Find out what it's like to have blubber! In this activity, you will make a blubber glove and learn what blubber does in the cold and the heat.

What You Need

- 2 1 litre freezer bags
- 3 cups (570 g) of solid vegetable shortening
- large serving spoon
- bowl of ice water (at least 20 cm deep)
- dry towel
- warm washcloth

Directions

1. Fill one freezer bag with vegetable shortening.
2. Turn the second bag inside out, place it inside the first bag, and press the edges of the two bags together. The pocket that is formed will be your "glove."
3. Gently knead the shortening to distribute it evenly between the bags.
4. Dip your bare hand into the ice water to feel how cold it is. Record your observations on page 19.
5. Dry your hand with the towel and place it inside the blubber glove. Dip your gloved hand in the water for one minute. Take your hand out of the water and remove the glove. Record your observations.
6. Soak the washcloth in hot water and wring it out. What does it feel like without the glove? Record your observations.
7. Place the warm washcloth inside the blubber glove. Feel the outside of the bag. Record your observations.

Make an **X** in the box that best describes the temperature of each item.

	Cold	Cool	Room Temperature	Warm	Hot
Water without glove					
Water with glove					
Washcloth without glove					
Washcloth inside glove					

What Did You Discover?

1. With your hand in the glove, did you feel the coldness of the water? How did it compare to putting your hand in the water without the glove?

____

____

____

____

2. Could you feel the heat of the washcloth from outside the glove?

____

____

____

3. Blubber prevents heat from escaping a polar bear's body. How would blubber affect a polar bear forced to live in a warm environment?

____

____

____

Switching Places

Skill:

Write narratives to develop real or imagined experiences or events.

What would happen if a polar bear and a panda switched places for a day? Write about a polar bear living in the forests of China and a panda living on the polar ice caps. Describe the challenges each bear would face.

TARGETING SCIENCE YEAR 5 © PASCAL PRESS ISBN: 9781925726541

First Nations people Perspectives - Structural Adaptations

Structural adaptations are the physical features of an organism that have evolved over time to help it to survive in its environment. For example, feathers, fur, teeth and claws help animals stay warm and find food. Spikes on the stems of plants and waxy coverings on leaves protect the plant and help it retain water.

First Nations people observed structural adaptations in plants and animals in their environments and incorporated them into a variety of weapons and tools to help them with their everyday lives. This is called **biomimicry.**

An example of this is the stingray barb. The barb (or stinger) is located on the stingray's tail. It has an extremely sharp point, often has backward serrations, and can also be venomous.

The barb is a defence mechanism that the stingray uses to defend itself when it is threatened by predators. When a stingray is faced with danger, it may strike the predator with this barb. The sharp tip helps to pierce the skin, the serrations ensure the barb does not fall out, and the venom causes pain and tissue damage.

First Nations people observed these adaptations and adopted these defence mechanisms, using their design and structure to make weapons for defence and hunting.

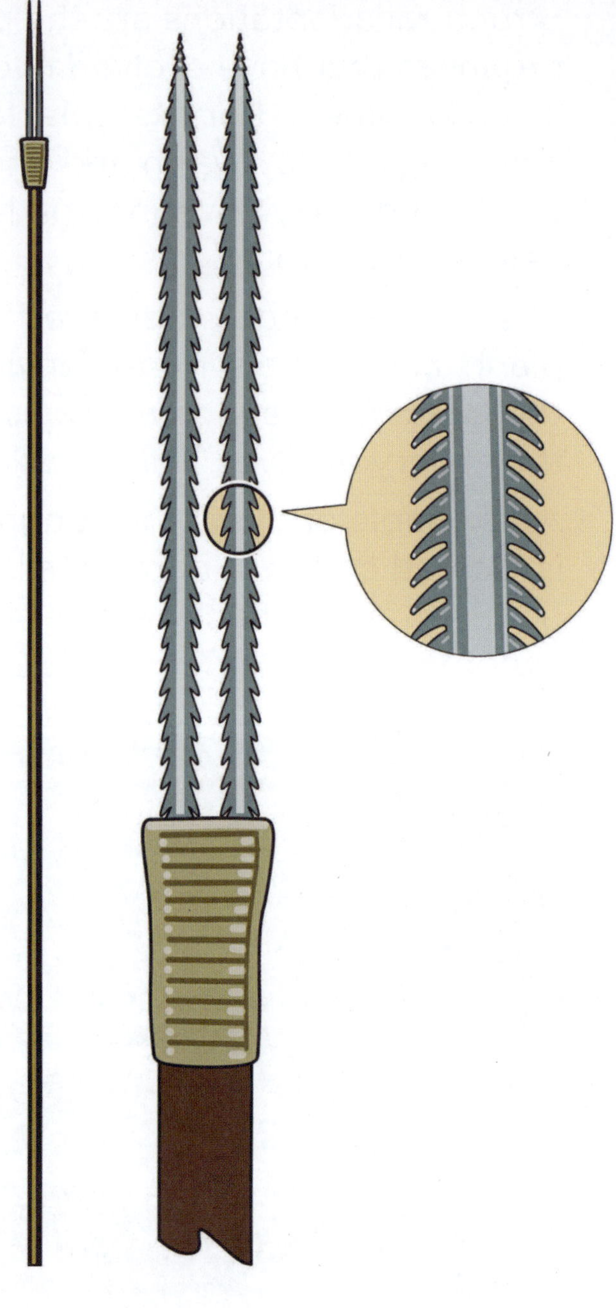

The Koko Tai'Yuri Peoples of the Kunjen Nation in Cape York made spear tips from stingray spines which shattered when they came into contact with something. The Gadigal People of the Eora nation, in the area now known as Sydney, attached multiple stingray barbs in a downward position on the spear shaft so that the serrations locked the tip of the spear into the flesh of fish and animals. In the Brisbane region, one or two stingray barbs were fastened to a fighting spear, using twine and beeswax to attach the barbs to the spear shaft.

First Nations people also observed that the mouth of a shark was a highly effective structural adaptation that was ideal for cutting through the flesh of seals and other ocean mammals.

Design a weapon or tool you think could have been made using the structural adaptations of a shark tooth.

Animals often live together in groups so they can help each other find food. Lions are a good example of this because they live in large families called **prides**. A pride might have 10 or 12 lions.

Define It!

carnivore: a meat eater

herd: a group of animals that keep together

prey: an animal hunted for food

pride: a group of lions

Because they are **carnivores**, or meat eaters, lions must hunt for their food. Most of their **prey** can run faster than lions do. So lions work together to catch their prey. One mother lion circles a **herd** of zebras and hides in the tall grass. Another lion sneaks up close to the herd and then runs at a zebra. The frightened zebra runs toward the hiding lion. The lion comes out of hiding, jumps on the zebra, and brings it to the ground. Everyone in the pride shares the meat.

Concepts:

Being part of a group helps animals obtain food.

Lions live in a group and hunt together.

Circle *true* or *false*.

1. Lions hunt because they are carnivores. **true** **false**
2. Lions work together when they hunt. **true** **false**
3. Most of their prey cannot run as fast as lions do. **true** **false**

Guarding the Group

Concepts:

Being part of a group helps animals defend themselves.

An elephant herd defends its members.

Define It!

attack: to move against with force

charge: to rush forward

rumbling: a long, low, heavy sound

trumpeting: a loud sound like a trumpet

African elephants travel in a herd. All of the elephants in the herd take care of each other. The leader of the herd is an old mother elephant. The other elephants follow her because she knows how to guard the family if danger appears.

Elephants talk to each other with sounds. Some sounds are so deep and **rumbling** that people cannot hear them. If danger comes near, the rumbling stops. Everyone is on guard. Mothers flap their ears to call their babies to them. The elephants circle around the baby elephants to guard them.

If a lion suddenly **attacks**, the elephants make **trumpeting** sounds and hit the ground with their trunks. The leader puts herself in front of the herd. She flaps her ears out to make herself look even larger. Then she lowers her head and **charges** the enemy in a big cloud of dust.

Write the missing words.

1. Elephants in a _______________ take care of each other.
2. Elephants make _______________ and _______________ sounds.
3. The leader guards the herd from _______________.

TARGETING SCIENCE YEAR 5 © PASCAL PRESS ISBN: 9781925726541

Define It!

flock: a group of birds together

habitat: a natural home

migrate: to move from one habitat to another

raft: a group of ducks on water

wedge: birds flying in a V-shape

Have you ever seen a **raft** of ducks swimming or a **wedge** of geese flying? Those are names for **flocks** of birds. Why do birds flock together? Some birds feed in a group because it is easier to find food. If one bird finds food, all can feed on it.

Sometimes birds flock together when their **habitat** changes. When the seasons change from summer to autumn, the berries, seeds, and bugs that birds eat are harder to find. So birds **migrate** to warmer places that have more food. Geese, ducks, and swans fly together in a **wedge**, or V-shape, when they migrate. They honk to let each other know where they are. Even at night or in cloudy skies, they can keep together.

1. Why do some birds migrate?

____

2. How does a wedge of geese talk to each other?

____

Concepts:

Being part of a group helps animals obtain food and cope with changes in their habitat.

Birds migrate in order to find a habitat with abundant food.

Migrating Monarchs

Monarch Butterflies are well known for the long distances they travel in different countries around the world.

In Summer, these large dark orange butterflies can be found mainly on the East coast and Southern West Coast of Australia, fluttering between milk weed plants, where they will lay their eggs and the tiny yellow and black striped caterpillars will grow and change.

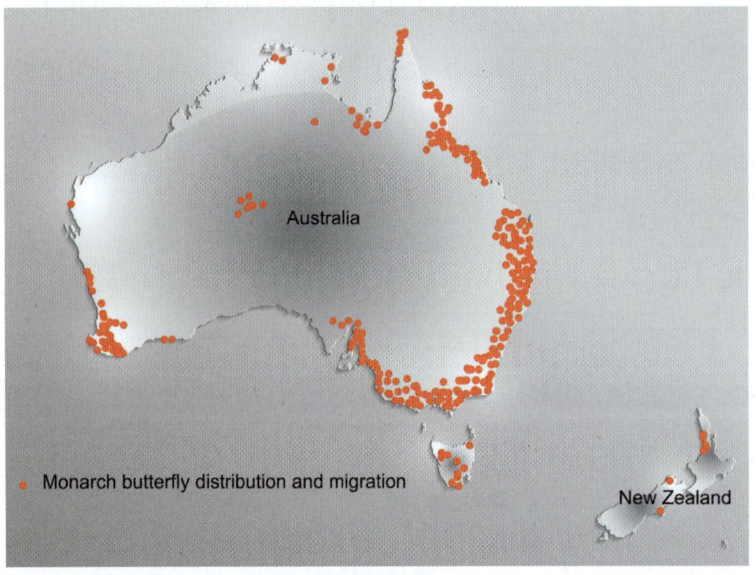

In Winter, the butterflies migrate (or move) closer to the coast, where it is warmer. If it is really cold, they gather together in large clusters and hang from the branches of particular trees until the weather warms up again.

1. Where do monarch butterflies migrate to in Winter months?

____

2. What do they do if it is really cold?

____

Look at this table:

Jan	Feb	Mar	Apr	May	June	July	Aug	Sep	Oct	Nov	Dec
Summer	Summer	Autumn	Autumn	Autumn	Winter	Winter	Winter	Spring	Spring	Spring	Summer

3. What months are monarch butterflies most likely to lay their eggs?

____

TARGETING SCIENCE YEAR 5 © PASCAL PRESS ISBN: 9781925726541

Adaptations Crossword Puzzle

Skill:

Apply content vocabulary.

Use the vocabulary words to complete the crossword puzzle.

habitat	carnivore	charge	pride	prey
trumpeting	rumbling	flock	migrate	

Across

2. an animal hunted for food
5. to rush forward
6. a long, low, heavy sound
7. to move from one habitat to another
9. a meat eater

Down

1. a group of lions
3. a natural home
4. a loud sound like a trumpet
8. a group of birds together

Adaptations

Looking at Animals

Skills:

Observe, collect, and record information using tools such as a camera and notebook.

1. What kinds of animals live where you live? Some may be pets. Some may be wild animals. Some may be in groups, and others may be alone.
2. Choose an animal or group of animals to watch closely. You can go for a walk with an adult to see animals in your neighbourhood. Or, you can watch your own pet at home. Take a camera with you.
3. Make a list of the things you see the animals doing. Are they eating? Looking for food? Building a home? Resting? Keeping away from danger? List all the things you see. Take some pictures.

Animal Name

____

Animal Actions

____

____

____

____

____

____

____

Animal Drawing or Photo

Adaptations

TARGETING SCIENCE YEAR 5 © PASCAL PRESS ISBN: 9781925726541

Animals Together

Skills:

Explain how being part of a group helps animals obtain food, defend themselves, and cope with changes.

Use information gained from words and illustrations to demonstrate understanding of text.

Choose an animal group from the list, or use one of your own ideas. Draw a picture of what the animal group does together. Then explain how being together in a group helps the animals to stay alive.

herd of elephants	pride of lions
flock of birds	swarm of butterflies

Draw

1. Name of animal group:

____

2. Explain what the group does that helps the animals stay alive.

How Young Animals Survive

Concepts:

Physical and behavioral adaptations allow organisms to survive.

Different organisms vary in how they look and function because they have different inherited information.

Offspring

Define It!

cub: a young bear

female: able to lay eggs or give birth to young

offspring: the young of an animal

sow: an adult female bear

survive: to stay alive

Animal babies are called **offspring**. The offspring of some animals must **survive** on their own. For example, the parents of a baby turtle, or turtle hatchling, are not around to keep it safe from bigger animals. **Female** turtles lay hundreds of eggs, but only some offspring will survive.

A black bear has fewer offspring. A female black bear, or **sow**, gives birth to two or three **cubs** in the middle of winter. The cubs are ready to leave their warm den in the spring. But the cubs stay close to their mother. She keeps them safe and teaches them to climb a tree quickly to get out of danger. The sow shows her cubs which plants, berries, and insects to eat. When the cubs are one and a half years old, they leave their mother to survive on their own.

Write the answer to each question.

1. Many turtle hatchlings do not survive. Why? ____________________

____

2. How long do black bear cubs live with their mother? ____________________

TARGETING SCIENCE YEAR 5 © PASCAL PRESS ISBN: 9781925726541

Define It!

cygnet: a young swan

inherit: to get from a parent

pen: an adult female swan

resemble: to be or look like

trait: a feature belonging to a living thing

Young animals grow up to **resemble**, or look like, their parents. Do you know the fairy tale called "The Ugly Duckling"? In that story, a baby hatches from an egg, but it doesn't look like the mother duck. Later in the story, the baby grows into a beautiful swan. A swan's egg had somehow gotten into a duck's nest! Ducklings grow into ducks and **cygnets** grow into swans.

Offspring **inherit** some of their **traits**, or looks, from each of their parents. Some traits appear as the young animal grows. Compare the traits of the cygnets and the **pen** (mother swan) in this picture.

Write the missing words.

1. The cygnets have fluffy grey feathers. The pen has smooth ______________ feathers.

2. The cygnets have a brown beak. The pen has an ______________ beak.

Concepts:

Different organisms vary in how they look and function because they have different inherited information.

Some inherited traits appear as an animal grows.

Review the Words

Skill:

Apply content vocabulary in context sentences.

Read each clue and write the missing word on the lines.

trait cygnet tadpole cub hind

1. A young frog with gills and a tail is a ___ ___ ___ ___ ___ ___ ___.

2. The offspring of a swan is a ___ ___ ___ ___ ___ ___.

3. A frog's ___ ___ ___ ___ legs grow before its front legs do.

4. A colourful beak is a ___ ___ ___ ___ ___ of some adult swans.

5. A black bear ___ ___ ___ stays with its mother for one and a half years.

Write the letters from the yellow boxes to answer the riddle.

Science Riddle

What lives in a little house and must break through the wall to go out?

a b___b___ c___ ___ ___k

Offspring

TARGETING SCIENCE YEAR 5 © PASCAL PRESS ISBN: 9781925726541

My Traits, Your Traits

Skills:

Gather and record scientific data from observation.

All living things pass on traits to their offspring. You may share traits with other people in your family. Compare your traits with a family member's. Circle one answer in each column.

Trait	Yours	Mine
Hair Colour	dark or light	dark or light
Hair Type	curly or straight	curly or straight
Eye Colour	dark or light	dark or light
Handed	left or right	left or right
Freckles	yes or no	yes or no
Dimples	yes or no	yes or no
Tongue Roll in a U-Shape	yes or no	yes or no

Shaping Our Planet's Surface

Concept:

Weathering and erosion help to shape our planet's surface.

https://clickv.ie/w/80gx

Use this QR code to access a video on this topic.

Define It!

erosion: the moving of rocks and soil by water, wind, ice, or gravity

gravity: a force that pulls objects toward the centre of Earth

landform: a natural feature of Earth's surface

weathering: the breaking down or wearing away of rocks by water or wind

Earth's surface is made up of many **landforms**. These landforms include mountains, valleys, islands, and canyons. The landforms on Earth have been shaped and reshaped by natural forces. Two of the most important forces in shaping our planet's surface are **weathering** and **erosion**.

Weathering is the breaking down or wearing away of rocks by water or wind. Have you ever picked up a very smooth rock on the beach? The smoothness of the rock is a result of the weathering waves of the ocean and the wind blowing on the beach. Erosion is the moving of rocks and soil by water, wind, ice, or **gravity**. High waves on a beach can erode sand dunes, carrying the sand back into the ocean.

Name one example of weathering and one example of erosion from the text.

1. weathering: ________________________________

2. erosion: ________________________________

TARGETING SCIENCE YEAR 5 © PASCAL PRESS ISBN: 9781925726541

Concept: Over millions of years, layers of sediment become layers of rock.

Define It!

deposit: to put or set something down in a specific place

sediment: very small pieces of sand and minerals set down by water, wind, or ice

spectacular: beautiful in an eye-catching way

One of Earth's most **spectacular** natural features is the 1.6 km deep Grand Canyon in the United States of America. It is also one of the best examples of weathering and erosion. Visitors can look from the rim of the canyon to see the Colorado River far below. The walls of the canyon have many layers of different kinds of rock. Some of the rocks are as much as two billion years old!

Over millions of years, small pieces of sand called **sediment** were **deposited** in the area where the Grand Canyon formed. As new layers of sediment were deposited, the older layers were pressed down. Over time, they became solid layers of rock. But it wasn't until five or six million years ago that weathering and erosion began cutting into the rock to form the canyon.

Answer the questions.

1. How deep is the Grand Canyon? ______________________

2. About when did the Grand Canyon begin to form? ______________________

Soil Profile

Soil makes layers or horizons during its formation. Floods, rain, wind and both geological (land) and meteorological (weather) events can cause soil to move and settle from one place to another. These layers or horizons are known as the soil profile.

A soil profile shows us a side-on view of these layers, and tells us a lot about the history of the soil in that area.

The layers of soil can easily be identified by the soil colour and size of soil particles. Layers in soil can also differ in their texture, structure, and thickness, as well as their chemical composition.

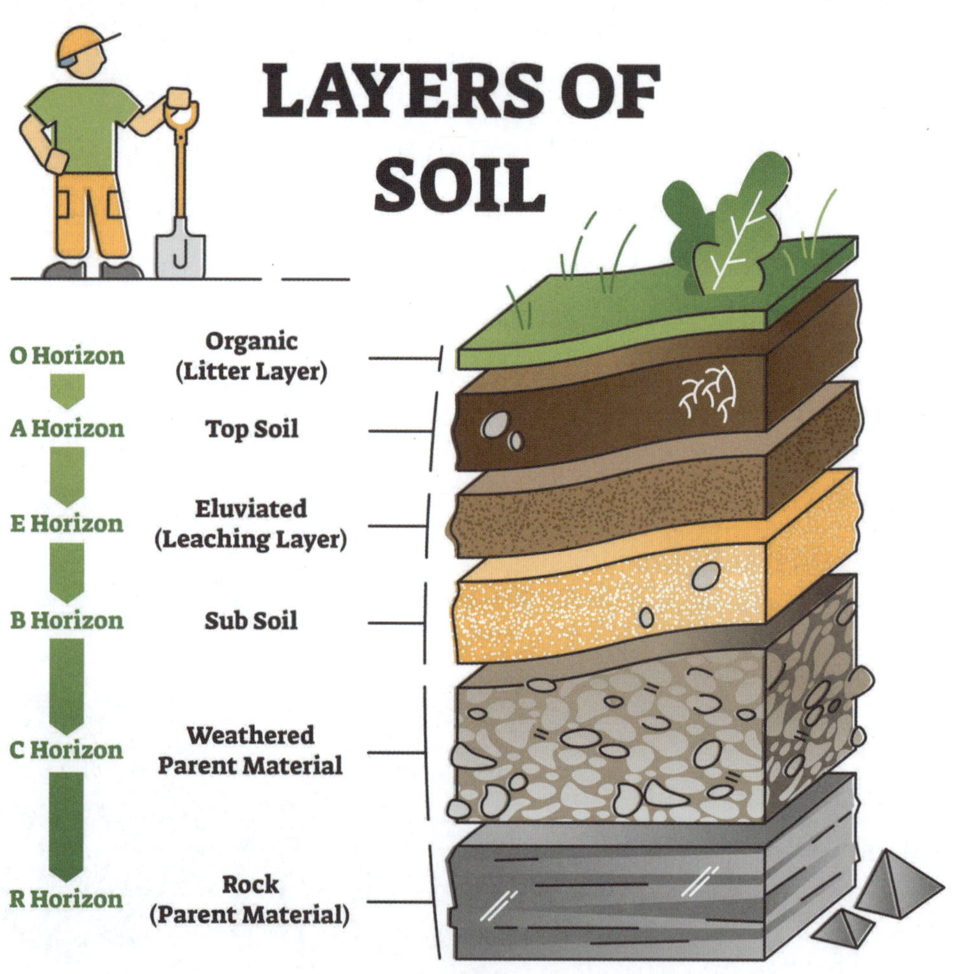

1. Can you think of some meteorological and geological events that could cause large amounts of soil to be shifted from one place to another?

____

2. What are some factors that would impact on the amount of erosion that could take place?

____

TARGETING SCIENCE YEAR 5 © PASCAL PRESS ISBN: 9781925726541

Shaping the Grand Canyon

Concept:
Erosion and weathering helped to form the Grand Canyon.

Define It!

channels: cuts in the ground made by moving water such as a river or stream

expand: to become larger

sloping: slanting; on an angle

Five or six million years ago, the Grand Canyon began to take shape. When rain fell, water ran down the **sloping** land of the Rocky Mountains. This eroded the soil, making **channels**. Over time, the channels became the path for the Colorado River. Over millions of years, the Colorado River kept eroding the soil and carving out the canyon.

Weathering also helped form the canyon. Rainwater ran into cracks in the rocks and froze in the winter. When the water froze, it **expanded** and pushed the rocks apart. Gravity caused sections of the canyon wall to fall, making the canyon wider. Wind also shaped the canyon. Bits of sand, blown by wind, chipped away at the canyon walls and weathered the rock. All these forces are at work even today, and they continue to change the canyon.

5 million years ago

Today

Circle all of the things that helped to shape the Grand Canyon.

gravity	**earthquakes**	**wind**
rainwater	**mudslides**	**ice**

Weathering & Erosion

How the Canyon Became Grand

Skills:

Write an opinion piece supporting a point of view with reasons.

Draw illustrations that show scientific concepts.

Think about how the land in the area of the Grand Canyon probably looked five million years ago. Think about how it looks now. In what ways has it changed? Explain what forces helped to make those changes.

5 million years ago

Today

____

____

____

____

____

What do you think the Grand Canyon will look like five million years in the future? Draw it.

Draw It!

Skills:

Interpret and identify information in photographs.

Look at the pictures of the rocks. On the lines below, describe what each rock looks like. Write down their colours, textures, and shapes. Then tell which one you think has been more weathered and why.

A

B

Which rock looks more weathered? Why?

Weathering/Erosion Crossword Puzzle

Use the vocabulary words to complete the crossword puzzle.

spectacular	landform	channels	sloping
weathering	sediment	erosion	expand

Across

3. a natural feature of Earth's surface
4. cuts in the ground made by moving water
8. the moving of rocks and soil by water, wind, ice, or gravity

Down

1. slanting; on an angle
2. the breaking down or wearing away of rocks by water or wind
5. to become larger
6. beautiful in an eye-catching way
7. very small pieces of sand and minerals set down by water, wind, or ice

TARGETING SCIENCE YEAR 5 © PASCAL PRESS ISBN: 9781925726541

Home Erosion

Skills: Conduct experiments and draw conclusions about the results.

Have you ever made a sand castle on the beach? Chances are, it wasn't very long before the castle was knocked down. What might make your sand castle stronger? And how does erosion affect sand castles or other structures made of sand and soil? Find out in this experiment.

What You Need

- sand
- soil
- 2 buckets
- 4 paper cups
- portable fan (with batteries)
- garden hose with sprayer

Directions

1. Before you start the experiment, look at the sand and soil. How do they feel? Do the sand and soil stick together easily, or do they fall apart? Record your observations.

Chart 1	Description
Sand, before experiment	
Soil, before experiment	

2. Make a prediction. Which structure do you think will stay standing longer—one made of dry sand or one made of dry soil? Explain your answer.

____

____

3. Pack one paper cup with dry sand and turn it upside down on the ground in order to make a structure. Repeat the same thing with another paper cup and the dry soil.

Weathering & Erosion

4. Create mud by mixing some of the soil with water in one bucket. In the other bucket, create wet sand by mixing some of the sand with just enough water to make it stick together.

5. Now make two more structures, one with mud and one with wet sand.

6. Blow on each of the structures one at a time, using the fan. Record what you observe in Chart 2.

7. Reusing the paper cups, make four more structures by repeating Steps 3, 4, and 5.

8. Holding the hose, stand about half a metre from your structures. Spray all the structures for about 30 seconds and then turn off the sprayer. Record what you observe in Chart 2.

Chart 2	Dry Soil	Dry Sand	Mud	Wet Sand
Wind (fan)				
Water (hose)				

What Did You Discover?

1. Which structure fell down first to the fan? Which stood up the best?

____

____

2. Which structure stood up best against the water?

____

 TARGETING SCIENCE YEAR 5 © PASCAL PRESS ISBN: 9781925726541

The Geosphere and the Grand Canyon

Earth's surface is very diverse. It is made up of landforms such as mountains, valleys, islands, and canyons. These landforms are part of the planet's **geosphere**. The geosphere contains all of Earth's **landforms**, as well as the rocks and minerals of its interior. The geosphere is constantly being shaped and reshaped by natural processes.

We can see natural geological processes at work at the Grand Canyon in America. The Grand Canyon began forming five or six million years ago, after forces within Earth pushed up the land and formed the Rocky Mountains. When rain fell, water ran down the mountains and began to **erode** the soil, making **channels**. These channels eventually became the path for the Colorado River. Over millions of years, the Colorado River continued to erode the soil and carve out the canyon.

Define It!

channels: cuts in the ground made by moving water such as a river or stream

erode: to move rocks and soil by water, wind, ice, or gravity

geosphere: the solid part of Earth, including its interior, rocks, and surface

landform: a natural feature of Earth's surface

Concepts:

The geosphere is constantly being shaped and reshaped by natural processes.

Number the events that formed the Grand Canyon in order.

_____ Forces pushed up the land to form mountains.

_____ As the channels got bigger and deeper, a river formed.

_____ Eventually, a deep canyon was formed with a river at the bottom.

_____ Water running off the mountains cut channels into the ground.

Weathering and Erosion

Concepts:

The Grand Canyon was formed by weathering and erosion.

Define It!

erosion: the moving of rocks and soil by water, wind, ice, or gravity

weathering: the breaking down or wearing away of rocks by water or wind

The Grand Canyon has been shaped over millions of years by two forces: **weathering** and **erosion**. Weathering is when rock is worn away or broken down. Erosion is when rock or earth is carried away. Erosion from the running water of the Colorado River was the major force that formed the canyon. However, weathering from rainwater also played a role.

Water from rain seeped into cracks in the canyon rocks and froze in the winter. When the water froze, it expanded, or got bigger, and pushed the rocks apart. This weathering process is called frost wedging. The pull of gravity then caused sections of the canyon wall to fall into the river, making the canyon wider. Wind also shaped the Grand Canyon. Bits of sand blown by wind chipped away at the canyon walls and weathered the rock. All of these forces are always at work, continually changing the canyon.

List the ways that erosion and weathering helped create the Grand Canyon.

1. ______________________________

2. ______________________________

3. ______________________________

4. ______________________________

TARGETING SCIENCE YEAR 5 © PASCAL PRESS ISBN: 9781925726541

People Impact Geological Processes

Define It!

deposit: to lay down or leave behind by a natural process

habitat: the natural home of an animal, plant, or other organism

sandbar: a raised area of sand with a top that is near or just above the surface of the water

Weathering and erosion can happen over long periods of time or in quick events such as avalanches or mudslides. At the Grand Canyon, the Colorado River can sometimes flood, carrying away rocks and sand that block parts of the river. Floodwaters **deposit** sand along riverbanks and create **sandbars**. These sandbars often become **habitats** for local plant and animal life.

In 1963, people built a dam on the Colorado River called the Glen Canyon Dam. This meant that the natural flooding stopped. Scientists later realised that without flooding, the plants and animals living in the Grand Canyon suffered. Now the dam is occasionally opened to release water and help preserve the natural habitats in the Grand Canyon.

Answer the questions.

1. How does the flooding of the Colorado River help erode the Grand Canyon?

____

____

2. How did the Glen Canyon Dam impact the process of erosion?

____

____

3. How did the Glen Canyon Dam impact the wildlife at the Grand Canyon?

____

____

Concepts: People can impact geological processes.

Weathering & Erosion

Weathering and Erosion at the Canyon

Skill:

Interpret information from graphic images.

Study the drawings of the Grand Canyon. Which types of weathering and erosion might be taking place? Describe what the pictures show.

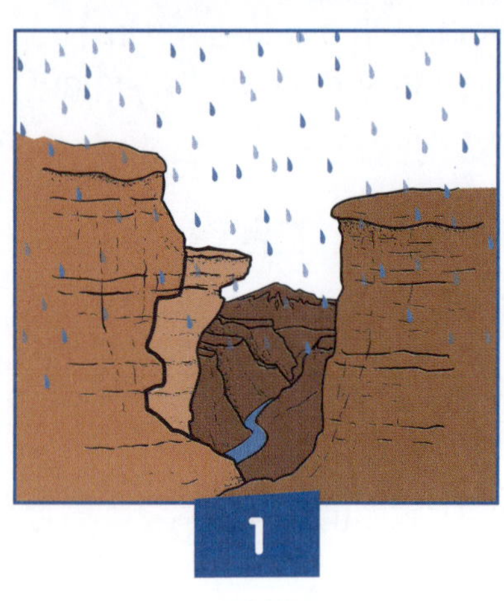

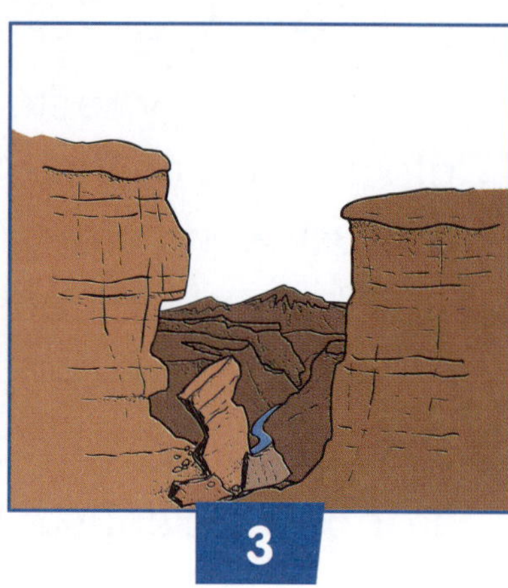

1. ____

____

____

2. ____

____

____

3. ____

____

____

TARGETING SCIENCE YEAR 5 © PASCAL PRESS ISBN: 9781925726541

Ways to Stop Erosion

Soil erosion is the moving of rocks and soil by water, wind, ice or gravity. Erosion is a serious problem. It causes pollution and the dumping of sediment into streams and rivers, clogging waterways and causing declines in fish and other species. Land degraded by erosion is also often less able to hold onto water, which can worsen flooding.

Serious erosion can cause landslides, sometimes even causing houses and buildings to slide down slopes.

There are many ways people have devised to try to prevent erosion.

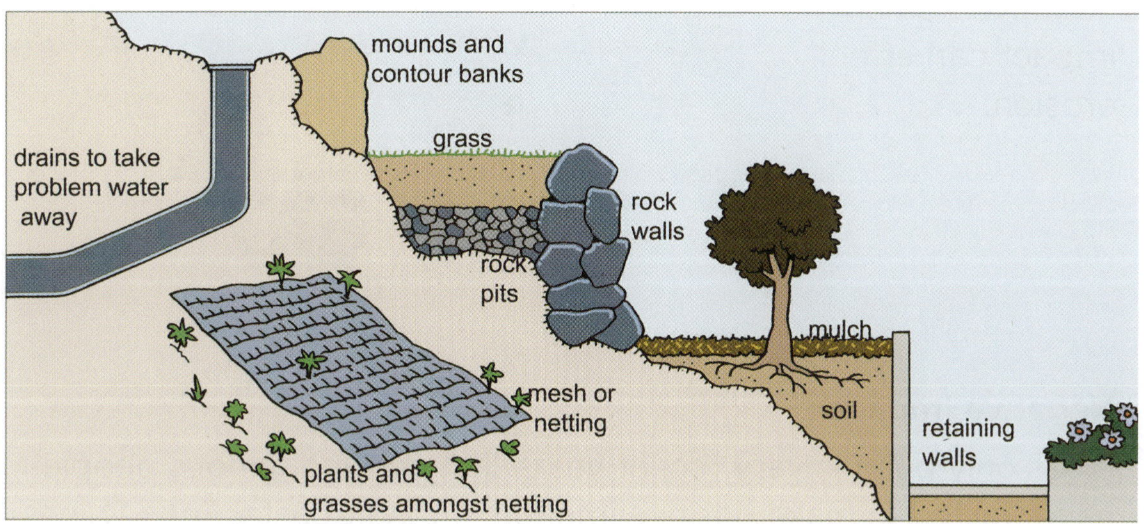

Use some of these methods to try to stop the erosion on this slope.

First Nations Perspectives on Sand Dune Erosion

The Stockton Bright Coastal Sand Dunes in New South Wales is an amazing example of how positive and negative human impact can effect erosion.

Negative Impacts:

It has only been in very recent years that non-Indigenous people have started to make positive impacts on the Coastal Sand Dunes at Stockton Bight. Disruption of sand flow, erosion from recreation such as motorbikes and four-wheel drives, cattle grazing, introduced weeds and pests, building developments and sand mining have all had very negative effects on the sand dunes, which has negatively impacted the flora and fauna of the area.

In recent years, non-Indigenous managers have started to eradicate the introduced species such as Bitou Bush, reduce access to certain parts of the Dunes, restabilise the foredune , and conserve and protect threatened species.

 TARGETING SCIENCE YEAR 5 © PASCAL PRESS ISBN: 9781925726541

Positive Impacts:

Vegetation plays two very important roles on coastal sand dunes. It helps stabilise the sand dunes by trapping and stabilising the grains of sand, and it creates a habitat for animals to live and breed.

In the Stockton Bight Coastal Sand Dunes, The Worimi people, through a practise called "fire stick farming", have actively encouraged vegetation growth. Many plant species in this area depend on fire for their seeds to germinate and for new plant growth.

By deliberately lighting fires, First Nations peoples have contributed to the growth of the thick forests and woodlands that were found in the secondary and tertiary vegetation zones. By providing an extensive habitat many other animal species have been able to thrive, thus making the local First Nations people partially responsible for the biodiversity along Stockton Bight.

1. Do you think people should be able to build houses in areas like this, or should nature come first? Can you write some for and against statements about this topic?

____

____

____

____

____

____

____

Vocabulary Practise

Skill:
Apply content vocabulary.

Select from the list of vocabulary words to complete the crossword puzzle.

channels	erode	sandbar	deposit	weathering
habitat	landform	erosion	geosphere	

Across

3. an animal's natural home
4. a raised area of sand in the water
6. the solid part of Earth
7. the moving of rocks and soil by water, wind, ice, or gravity

Down

1. the breaking down or wearing away of rocks
2. a natural feature of Earth's surface
5. cuts in the land made by moving water

Weathering & Erosion

TARGETING SCIENCE YEAR 5 © PASCAL PRESS ISBN: 9781925726541

Effects of Erosion

Skills:

Conduct experiments, record data, and analyse results.

See the effects of erosion in this simple experiment involving dirt and water.

What You Need

- dirt
- water
- watering can
- magnifying glass
- 2 large disposable aluminium baking pans
- books that equal a total height of 5 cm (or use any other object that could be used to prop something up)
- pointy scissors or knife

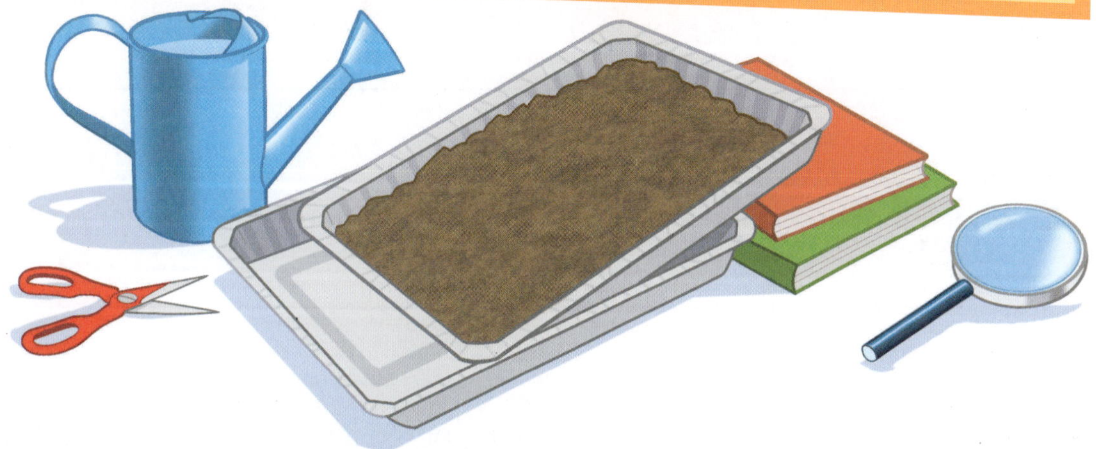

Directions

1. Fill one pan with a layer of dirt about 5 x 8 cm deep. Smooth out the dirt so that it is even.
2. Use the magnifying glass to examine the soil. Record your observations on page 50.
3. Use scissors to punch 6 small holes in the bottom of one end of the dirt-filled pan.
4. Place the second pan under the end of the dirt-filled pan (where the holes are located) and prop up the other end with books as shown. The pan should be sloped so that one end is about 5 cm higher than the end with the holes.
5. Pour water from the watering can onto the raised end of the dirt-filled pan. Record your observations.

	Observations
Dry dirt	
Wet dirt	

What Did You Discover?

1. What was the soil made of? How did it feel and look before you poured water on it?

2. What happened to the surface of the dirt when you poured water on it?

3. What happened to the water after you poured it onto the dirt?

4. Do you think it would make a difference if the soil started out wet or dry? Explain your answer.

5. What do you think would happen if you created a steeper slope for the dirt-filled pan?

 TARGETING SCIENCE YEAR 5 © PASCAL PRESS ISBN: 9781925726541

Skill:
Present information through graphic images and text.

Find examples of geological features that have been created by weathering and erosion. Collect pictures from magazines or websites and arrange them on a bulletin board or large sheet of paper. Below each picture, write a description of the structure, its location, and one or two sentences on how it was formed.

Example features include:

- Upper and Lower Antelope Slot Canyons, Arizona
- Gumdrop Hills, Guilin, China
- Karst, Tsingy de Bemaraha, Madagascar
- Bryce Canyon, Utah
- The Twelve Apostles, Victoria, Australia
- Hoodoos, Banfora, Burkina Faso
- Phu Phrabat Historical Park, Thailand
- Horseshoe Bend, Arizona

Horseshoe Bend

Antelope Slot Canyons

About Floods

Concepts:

Floods are a natural hazard that results from natural processes.

A flood can cause harm to people and property.

https://clickv.ie/w/2zgx

Use this QR code to access a video on this topic.

Define It!

banks: land at the edge of a river

cycle: a pattern of events that happen in the same order

natural hazard: an event in nature that causes harm to people and property

property: something that is owned

When a body of water covers an area that is usually dry land, that is a flood. Areas near rivers, lakes, and oceans can flood. River floods are the most common. Many rivers have a **cycle** of dry times and flooding. Rivers and lakes may flow over their **banks** when there is more rain than usual, or when snow and ice melt suddenly. Cyclones can bring floods to areas along ocean coasts.

Although it can sometimes do good things, such as bring new soil to an area, a flood is a **natural hazard**. Farms and crops may be destroyed by a flood. In an area where many people live, floods can be a danger to people's lives. More lives are lost in floods than in any other kind of weather. Floods destroy **property**, and building over again after a flood can cost billions of dollars.

sittitap / Shutterstock.com

Answer the questions.

1. What can happen to rivers when snow and ice suddenly melt?

____

2. What can happen to crops in a flood?

____

TARGETING SCIENCE YEAR 5 © PASCAL PRESS ISBN: 9781925726541

Concepts:

Floods occurred in ancient times.

People cannot eliminate natural hazards.

Natural hazards such as floods can also have benefits.

Define It!

ancient: belonging to times long past

control: to hold back

dam: a barrier to hold back water

nutrients: elements in soil that are needed by plants

silt: fine sand or clay carried by water

The Nile River in Africa is the longest river in the world. In **ancient** times in Egypt, the Nile River flooded every year. To the people of ancient Egypt, the flooding Nile was a force of good. The river carried **silt**, or rich new soil. When the floodwaters went away, the rich soil remained. The new soil contained **nutrients** that were good for growing crops. The Nile was so important to the ancient people that their calendar was based on its cycle. The new year began in the middle of summer, when the river began to rise. Today, there is a huge **dam** on the Nile at Aswan, Egypt. The dam was completed in 1970. It **controls** floods in the rainy season and stores water to be used later.

Circle *true* or *false*.

1. The Nile River floods carried away old soil. **true** **false**
2. The people of ancient Egypt built a dam on the Nile. **true** **false**
3. The Nile is the world's longest river. **true** **false**

Flood Control

Concepts:

People cannot eliminate natural hazards such as floods, but they can take steps to reduce the impact.

A dam controls river flooding, stores water to be used later, and creates power.

Define It!

hectare: a unit of land

electric: a type of energy

engineer: a person who plans things to be built

reservoir: a large lake used as a water supply

turbine: a machine for making power with a wheel turned by water

Engineers plan dams that are built to control flooding. Hoover Dam stands on the Colorado River between Arizona and Nevada. Before Hoover Dam was built, melting snow in the mountains caused the Colorado River to flood every spring. Floods destroyed thousands of **hectares** of crops. But by summer, the river didn't hold enough water for farms. Hoover Dam was built to control the Colorado River. Behind the dam is a giant **reservoir** of water called Lake Mead. The water in the reservoir is used by farmers in Nevada, Arizona, and California to water crops. The water also goes to cities in southern California. Engineers control the water falling through the dam, and this water moves **turbines** that make **electric** power. The power is used in California, Nevada, and Arizona.

Complete the sentences.

1. Hoover Dam controls floods on the ______________________.
2. Lake Mead is a ______________ of water used by farms and cities.
3. Hoover Dam produces ______________________.

Floods

 TARGETING SCIENCE YEAR 5 © PASCAL PRESS ISBN: 9781925726541

Types of Dams

Skills:

Analyse and interpret information presented in text and illustrations.

Label photos that illustrate a variety of types of structures.

Engineers build dams in different shapes and with different materials. Some dams are made from earth and others from concrete. Concrete is sand, stones, cement, and water mixed together. Concrete hardens and becomes very strong. These dams are all found in Australia.

Read about each type of dam. Then label the pictures to tell which type of dam each one is.

Embankment Dam An embankment is a wall or bank of earth. The Lake Wivenhoe Dam in Queensland is an embankment dam.

Gravity Dam It is thick and made of concrete. Its huge weight holds back the water. The Somerset Dam in Queensland is a gravity dam.

Buttress Dam Tall supports called *buttresses* hold up the dam. The Lake Hume Dam in New South Wales is a buttress dam.

Arch Dam It has a curved shape and is often built between two walls of rock. The Gordon River Dam in Tasmania is an arch dam.

1. Gordon River Dam

____

2. Lake Hume Dam

____

3. Somerset Dam

____

4. Lake Wivenhoe Dam

____

Make a Working Waterwheel

Skills:

Follow a sequence of steps to construct a model.

Modify a design and analyse the results.

A waterwheel is a machine that changes the energy of falling water into power. The power of the waterwheel turns an axle, or rod. A dam uses the power of falling water to run a motor that makes electricity. You can make a waterwheel with power to lift a weight.

What You Need

- 2 or more sturdy paper plates
- scissors
- pencil
- 30 cm piece of wool
- large paper clip
- sink with running water

What You Do

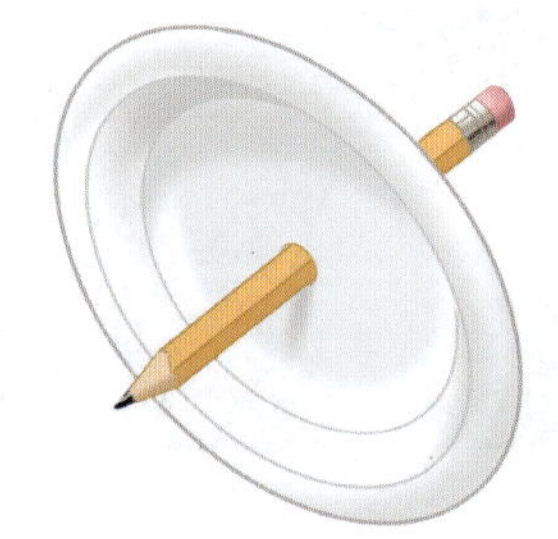

1. Ask an adult to poke a tiny hole in the centre of a paper plate, using scissors.
2. Poke the pencil halfway through the hole so the pencil fits snugly.
3. Cut slits about 5 cm apart around the edge of another plate. Bend them back and cut them off for flaps. Cut a small slit at the bottom of each flap as shown.

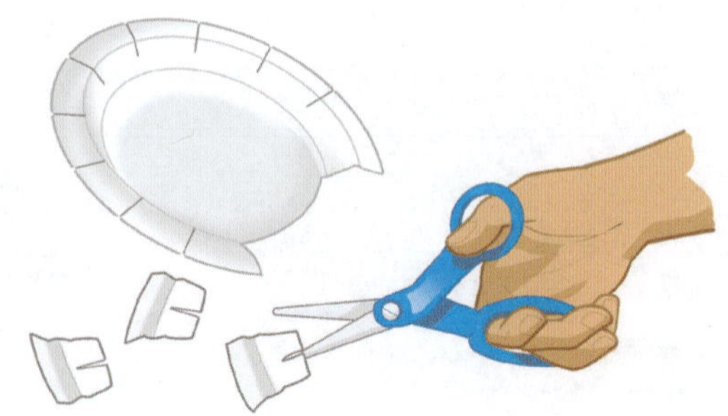

Floods

TARGETING SCIENCE YEAR 5 © PASCAL PRESS ISBN: 9781925726541

4. Cut 1.3 cm slits into the edge of the first paper plate, about 5 cm apart. Fit a flap into each slit on the paper plate, slits together as shown.

5. Tie the paper clip to one end of the wool. Decide where on the pencil to tie the other end of the yarn. Then test your idea.

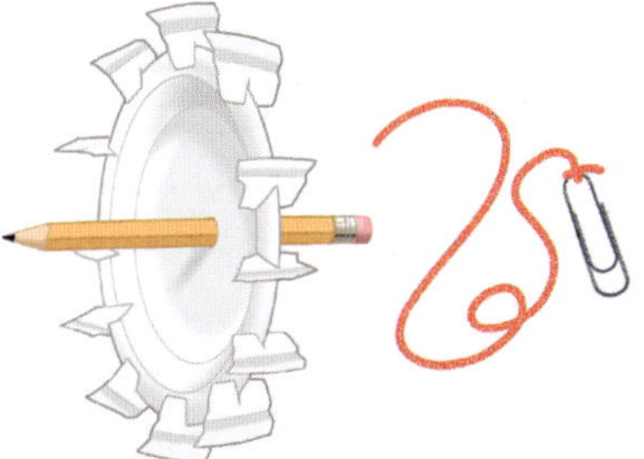

6. Lay the ends of the pencil on your open hands with palms up, so both the waterwheel and the pencil can turn. Let the wool and the paper clip drop down from the pencil. Hold the waterwheel over the sink in a stream of water as shown.

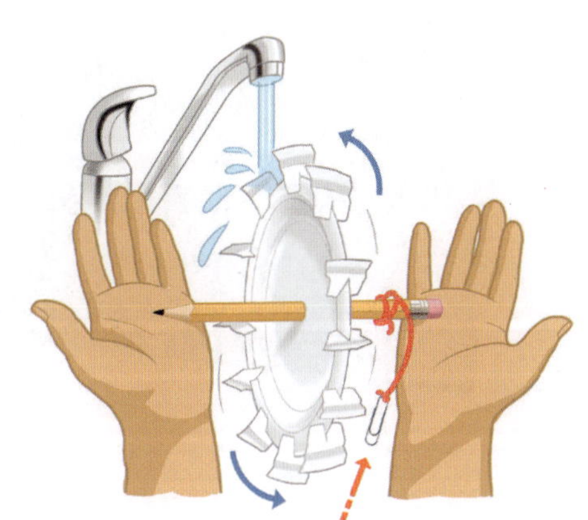

7. The wheel will turn, causing the pencil to spin. The wool should wind around the pencil, lifting the paper clip.

What Did You Discover?

Did your idea work?

____

What happens when you slide the wool to a different place on the pencil? Explain.

____

____

Can you make the wool unwind and lower the paper clip? How?

____

____

Thinking About Floods

Skill:

Write informative text to examine different views of the impact of a natural hazard.

There is more than one way to think about floods. Floods are a natural hazard, but the people of ancient Egypt considered them to be a good thing.

Write some ideas to support each way of thinking.

1. Floods are a natural hazard.

____

____

____

2. In ancient Egypt, floods were helpful.

____

____

____

 TARGETING SCIENCE YEAR 5 © PASCAL PRESS ISBN: 9781925726541

Light Sources

https://clickv.ie/w/z2gx

Use this QR code to access a video on this topic.

Light can come from both natural and artificial sources. For light to be considered natural, it must come from sources that occur naturally. Light from the sun, light from a campfire and even the glow of a glow worm in a cave are all examples of naturally occurring light sources.

Artificial light is emitted by man-made devices that would not occur naturally. Examples include lamps, televisions or the light from a mobile phone.

Label these light sources as natural or artificial:

Lamp	**Torch**	**Firefly**
Fireworks	**Stars**	**Traffic light**
Bushfire	**Lighthouse**	**Lightning**
Lightbulb	**Car Headlights**	**Sun**

Light Travels

Concepts:

Visible light waves allow us to see the world around us.

Light travels through empty space at 300,000 km per second.

Define It!

electromagnetic wave: a wave that travels at the speed of light

invisible: not able to be seen

laser: a tool that uses a strong beam of light

visible: able to be seen

The sun and other stars are always giving off energy in the form of **electromagnetic waves**. Light energy from the sun travels at 300,000 km per second through empty space to reach us on Earth. Other electromagnetic waves such as x-rays, radio waves, and microwaves are **invisible** to us. It is **visible** light that allows us to see the world around us. If you can see an object, that is because light is hitting the object and travelling to your eyes.

People have studied light for hundreds of years, but there is still a lot for scientists to wonder about. Scientists experiment with light to see how it behaves. This can help them to invent new uses for light. For example, a **laser** is a useful tool for doctors that uses a strong beam of light.

Answer the questions.

1. What is the only form of electromagnetic waves visible to us?

____

2. What is the speed of light travelling through empty space?

____

TARGETING SCIENCE YEAR 5 © PASCAL PRESS ISBN: 9781925726541

Define It!

opaque: not allowing light to pass through

ray: a thin beam of light

translucent: allowing some light to pass through, but not all

transparent: allowing light to pass through

Scientists sometimes talk about light as a **ray**, or a thin beam. Light rays move from place to place in a straight line. You see an object when light rays bounce off the object and travel to your eyes. Light rays pass through a **transparent** object, such as a window. **Translucent** objects let *some* light pass through, but not all. For example, sunglasses and tinted car windows are translucent. However, if a light ray hits an **opaque** object, such as a wall, it cannot pass through it. This is how shadows are made. When light can't pass through an object, a shadow is produced on the other side of the object.

Concepts:

Light passes through transparent objects.

Translucent objects let some but not all light pass through.

Light cannot pass through opaque objects.

Write *true* or *false*.

1. You can see through opaque objects. ____________
2. Light moves in a straight line. ____________
3. Transparent objects make shadows. ____________

Shadows

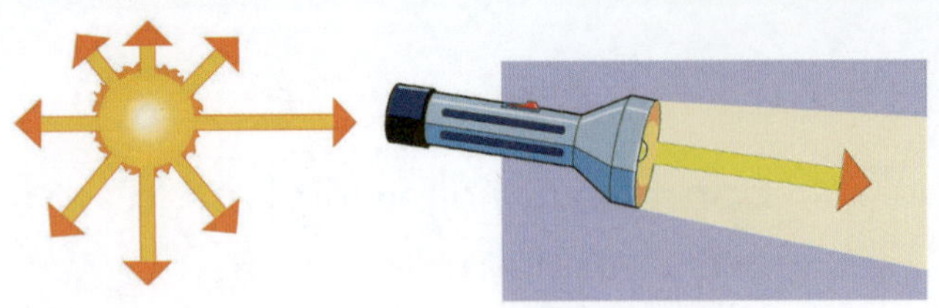

Light always travels in a **straight line** from its source. If there is something in the path of light rays, they are either blocked and the light is absorbed, reflected, and the light bounces back, or refracted, where the light changes direction.

Shadows are formed when the **light rays are blocked** from a naturally occurring or artificial light source.

If an object is **opaque**, and you cannot see through it, you will see a dark shadow.

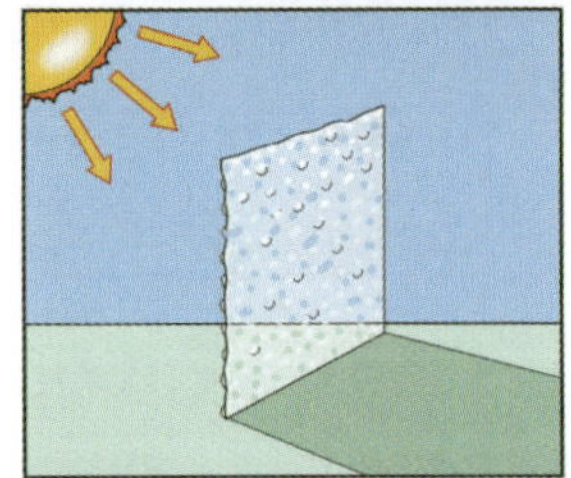

If an object is **translucent,** and you can see through it, but not clearly, you will get a lighter shadow.

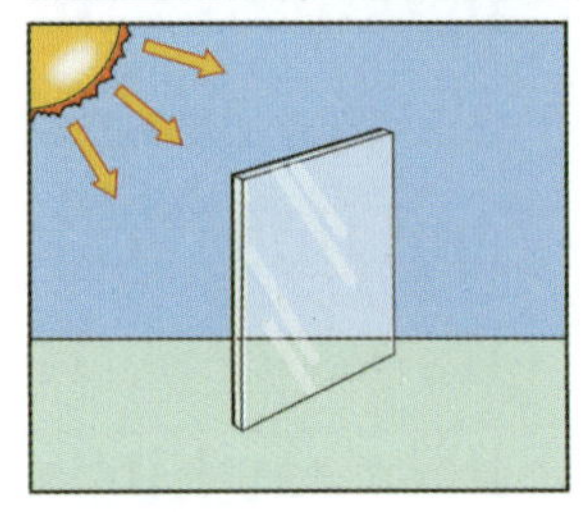

If an object is **transparent**, and you can see through it clearly, you will see a very faint shadow, or no shadow at all.

Predict whether the shadow from these objects will be dark, lighter or very faint. You could then find a torch and some objects of your own to test.

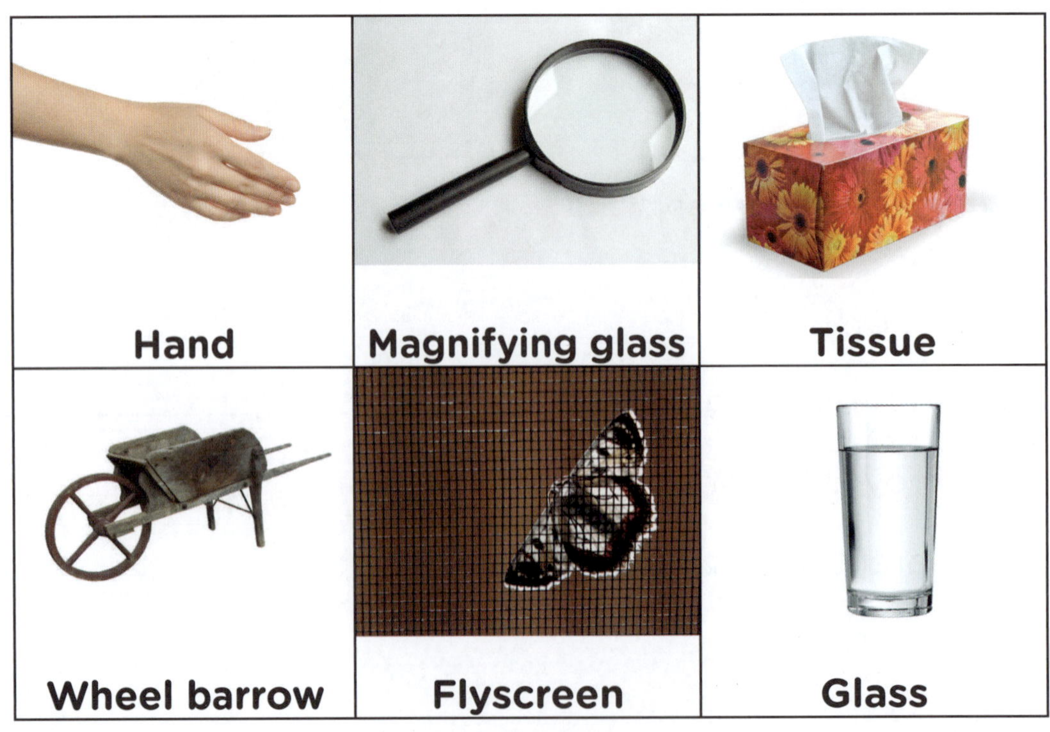

How Does Light Behave?

Define It!

absorb: to soak up

reflect: to send light back from a shiny surface

reflection: an image made by reflecting light

refract: to bend a light wave

Opaque objects, such as walls, **absorb** light that tries to pass through them. This light energy can heat the objects that absorb it. This is why you can feel heat if you touch a brick wall on a hot, sunny day.

Some opaque objects are shiny, such as a mirror or a piece of metal. Shiny objects **reflect** most of the energy that strikes them. A small amount gets absorbed. The light bounces off them and travels in a new direction. Your eyes gather that light, and you see your **reflection**.

Sometimes, light passes through a transparent object and bends before it reaches your eyes. For example, light cannot pass from air through water in a straight line. Water **refracts**, or bends, the light. This is what happens when you look at a drinking straw in a glass of water. The water refracts the light rays, and your eye is tricked into thinking the straw is bent.

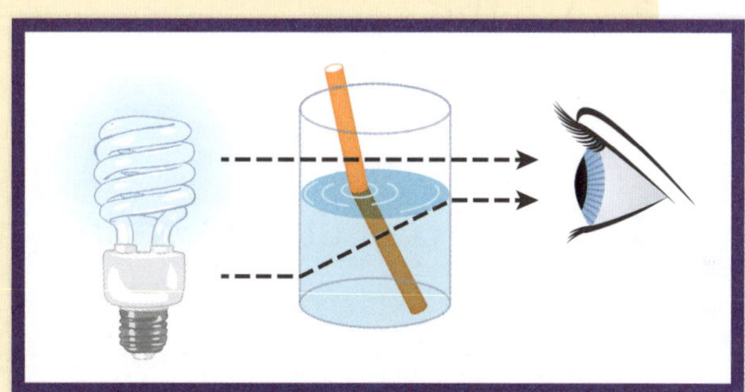

Complete the sentences.

1. An opaque object that is shiny will ______________________.
2. When light bends as it passes through water, we can say the water______________________.

Concepts:

Light is absorbed by opaque objects.

Shiny objects reflect light.

Some objects refract, or bend, light.

Light Energy

Light from the Sun

Skills:

Interpret and apply information gained from text and illustrations.

This picture shows some ways that sunlight affects us and our surroundings. Answer the questions to describe what you see.

1. Is the table *opaque, transparent,* or *translucent*?

____

2. Are the sunglasses *opaque, transparent,* or *translucent*?

____

3. Which objects are reflecting light? Name at least two.

____

____

4. Which objects are absorbing light? Name at least two.

____

____

5. Which objects show light being refracted? Name two.

____

____

TARGETING SCIENCE YEAR 5 © PASCAL PRESS ISBN: 9781925726541

A ray diagram is drawn to demonstrate the path that light takes when it travels from its source to a point on an object.

On the diagram, rays (lines with arrows) are drawn from where the light comes from (called the incident ray) to where it is reflected (reflected ray). People and other complex things are often represented as stick figures or arrows.

The boy in this picture is reading a book using a lamp as a source of light.

The light starts at the lamp and follows the line of incidence to a point on the book where it is reflected back following the line of reflection into the boys eyes, so he can see the writing.

Use a pencil and a ruler to draw ray diagrams for these two light sources. Make sure you label the line of incidence and the line of reflection.

Periscopes

A periscope is an ingenious instrument for seeing things from a hidden position, or even around a corner or over a fence! Most periscopes are made from a long tube with two mirrors. The mirrors are placed at each end of the tube at a 45-degree angle so that they face each other.

The light travels in from the source and hits the first mirror, which reflects and travels down the tube to the second mirror. The light is then reflected again into the eye of the person looking in.

Some periscopes use prisms instead of mirrors.

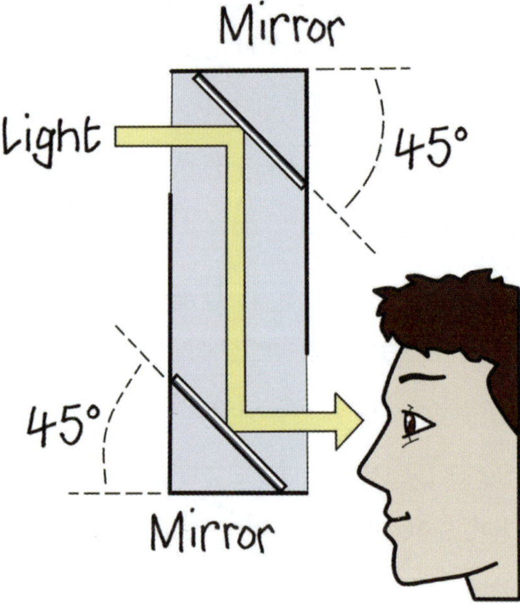

Make your own:

Light Energy

TARGETING SCIENCE YEAR 5 © PASCAL PRESS ISBN: 9781925726541

A First Nations Perspecitve on Refraction

When you place a small glass jar inside a larger one, and slowly fill the inside jar with cooking oil, then let it overflow to fill the larger jar, something magical happens. The glass jar on the inside seems to disappear!

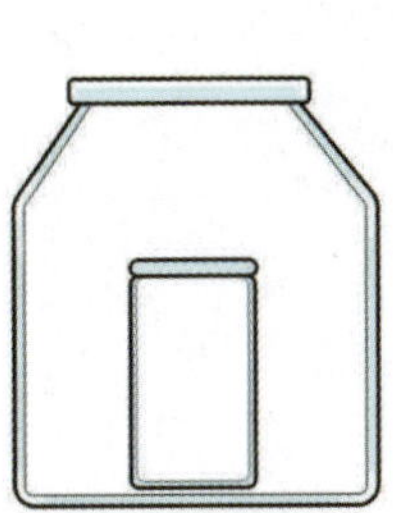

What happened?

Why did the glass seem to disappear?

The beaker is invisible in the oil because of the way light travels through the different mediums of air, glass and oil.

As light travels from one medium to another, it slows down and changes direction slightly. This is what we call **refraction**.

First Nations Australians were masters of understanding refraction. When they saw a fish in the water that they were trying to spear, they knew that they had to aim their spears closer than the fish appeared. As the light hit the water and slowed its speed, it made the fish appear further away than it actually was.

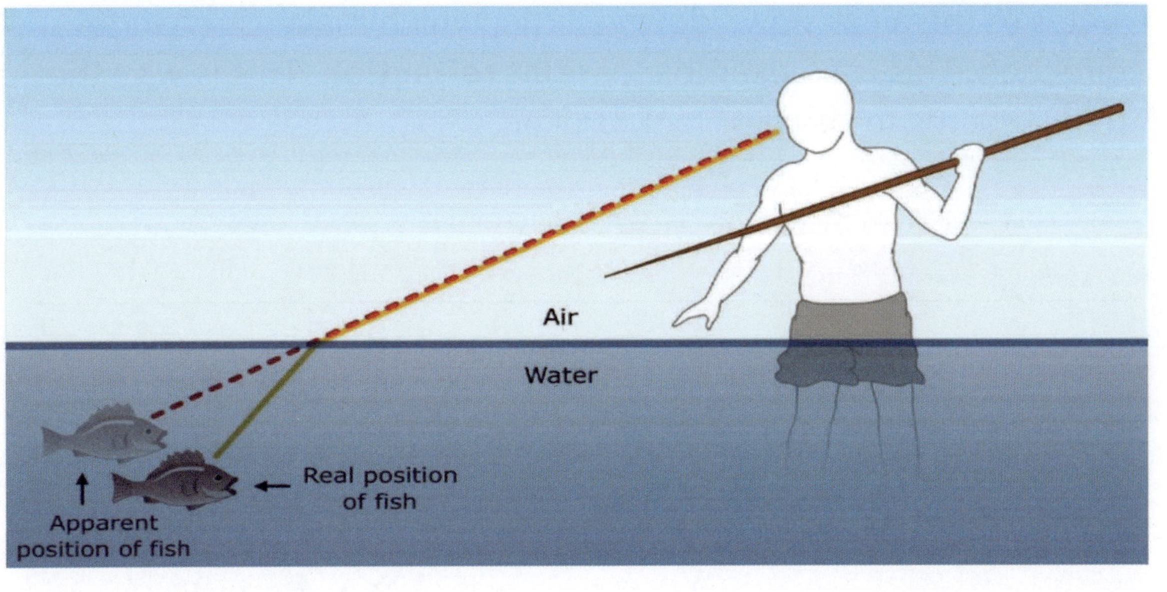

Light Energy

Refraction

Try this:

Find a large clear plastic cup or jar and fill it with water. Place a pencil or long thin stick in the water, leaning it against the edge.

Look at the pencil through the side of the cup

Look at the pencil from above.

Hold the pencil in the centre of the cup, perpendicular to the table.

What did you see when the pencil was leaning against the side of the cup?

What did you see when you looked at the pencil from above?

Is the pencil really bent or broken?

What did you observe when you stood the pencil up straight?

As the light reflects off the pencil and back to our eyes, it must pass through the water and then the air. As it passes from the water to the air, it slows down a little, which makes it change direction. This is an example of **light refraction**. This is why the pencil seems to bend or look broken.

____

____

____

____

Holographs

Holography uses the reflective properties of light to create an image that looks three dimensional. A special mirror splits light from a laser into two beams. One beam shines directly onto a piece of photographic film. The other shines on the object and reflects onto the film. The two overlapping beams create a pattern of closely spaced lines, which then appear to form a three-dimensional picture. You can make your own hologram using readily available materials.

Materials

- smart phone
- ruler
- sharp pencil
- piece of paper
- Stanley knife
- sticky tape
- clear, unscratched CD case

What You Do

Risks: Be careful not to cut yourself while using the Stanley knife!

1. Use the ruler to draw a trapezium on a piece of paper and then cut it out.
2. Trace the trapezium shape onto the clear cover of a CD case four times.
3. Use the Stanley knife to cut out the shapes, taking care not to scratch the case.
4. Assemble the four pieces into a shape like a pyramid without a top, using tape to hold them together.
5. Lay your smart phone flat on the table and play a video.
6. Put your plastic shape on top of the phone. Make sure the small opening is at the bottom.
7. Look at your phone through the shape's centre, and watch your hologram come to life!

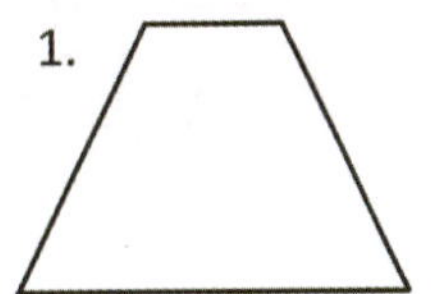

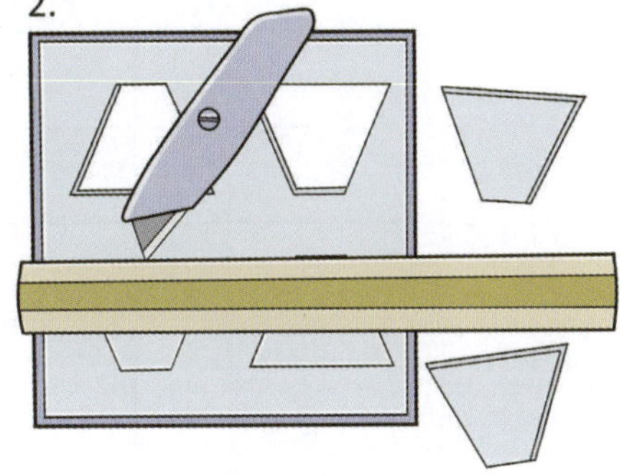

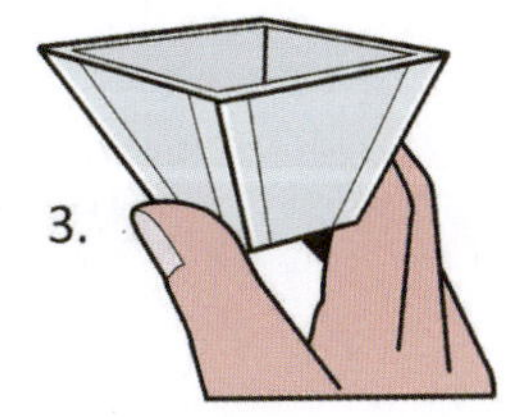

Light Energy Crossword Puzzle

Skill:

Apply content vocabulary.

Use the vocabulary words to complete the crossword puzzle.

transparent	laser	absorb	refract
translucent	opaque	reflect	ray

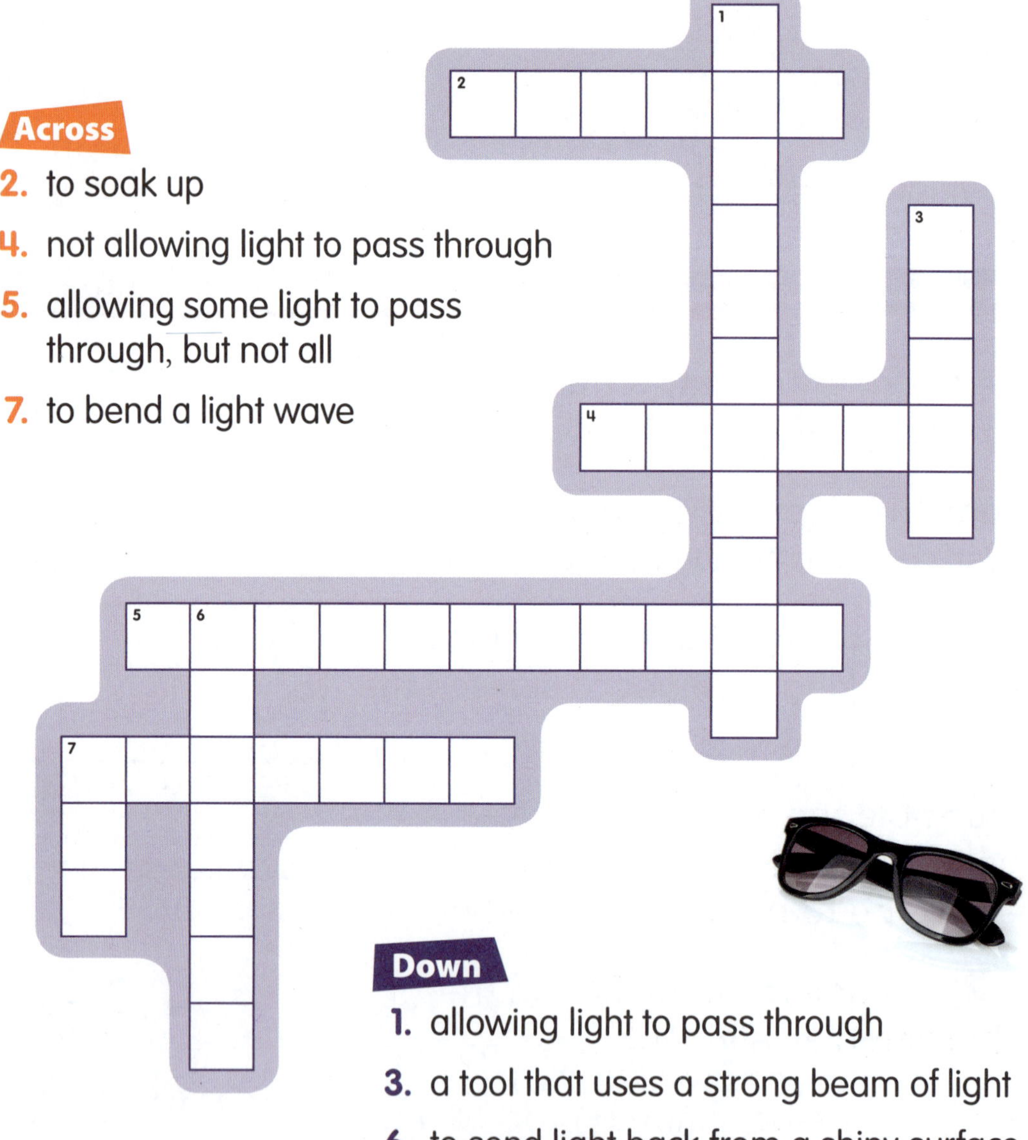

Across

2. to soak up
4. not allowing light to pass through
5. allowing some light to pass through, but not all
7. to bend a light wave

Down

1. allowing light to pass through
3. a tool that uses a strong beam of light
6. to send light back from a shiny surface
7. a thin beam of light

Light Energy

TARGETING SCIENCE YEAR 5 © PASCAL PRESS ISBN: 9781925726541

Sunlight and Heat Energy

On a hot and sunny summer day, do you feel cooler in the sun or the shade? Do you feel cooler in light-coloured clothing or dark-coloured clothing? Find out what this has to do with the energy in sunlight.

Skills:

Follow a sequence of directions to conduct a science investigation.

Record observations and interpret results.

What You Need

- 2 clear jars of the same size, with lids
- room-temperature water
- black construction paper
- white construction paper
- rubber bands or clear tape
- 2 thermometers
- clock

Directions

1. Fill both jars with the same amount of room-temperature water.
2. Measure and record the temperature of the water in each jar under "Start" on the chart on page 72. Also record the time.
3. Put the lids on the jars. Cover one jar and lid with black paper and the other jar and lid with white paper. Use rubber bands or tape to secure the paper.

Light Energy

	Time							
	Start							
Temperature in Black Jar (ºC)								
Temperature in White Jar (ºC)								

4. Place the jars in sunlight. Predict what will happen to the temperature of the water in the two jars.

____

5. Measure and record the water temperature and the time every 5 to 10 minutes. Move the jars if necessary to keep them in the sunlight.

6. What happened to the water temperature in the jars?

____

7. Which jar of water got warmer?

____

____

____

8. What heated the water in the jars?

____

9. From what you observed, which colour paper absorbs more light energy?

____

TARGETING SCIENCE YEAR 5 © PASCAL PRESS ISBN: 9781925726541

Wow!

Albert Einstein was a famous scientist who did "thought experiments". In one, he imagined what he would see if he were riding on a beam of light.

KENNY TONG / Shutterstock.com

Imagine your own thought experiment with light. Would you travel from the sun to Earth? Would you travel through something transparent or translucent? Would you be absorbed by something opaque? Tell your story. Show what you have learned about how light behaves.

Skills:

Apply scientific knowledge to writing a narrative about an imagined experience.

Use descriptive details.

Parts of the Eye

Concepts:

The eyes take in light and send signals to the brain, which tells us what we are looking at.

The parts of the eye we can see are the sclera, iris, and pupil.

Define It!

iris: the coloured part of the eye that controls the amount of light that can enter

pupil: the dark circle in the centre of the iris where light enters the eye

sclera: the white part of the eye that controls movement

spherical: shaped like a ball or globe

Have you ever wondered how you are able to see? Our eyes are the organs that control our sense of sight. They take in light and send signals to the brain. The brain then tells us what we are seeing.

Even though eyes may appear circular or almond-shaped, they are really **spherical**. The parts of the eye that we can see are the **sclera**, **iris**, and **pupil**. The sclera is the white part of the eye. It contains muscles that control the eye's movement. The iris is the coloured part of the eye. It controls the amount of light that can enter the eye. The pupil is the dark circle in the centre of the iris. Light enters the eye through the pupil.

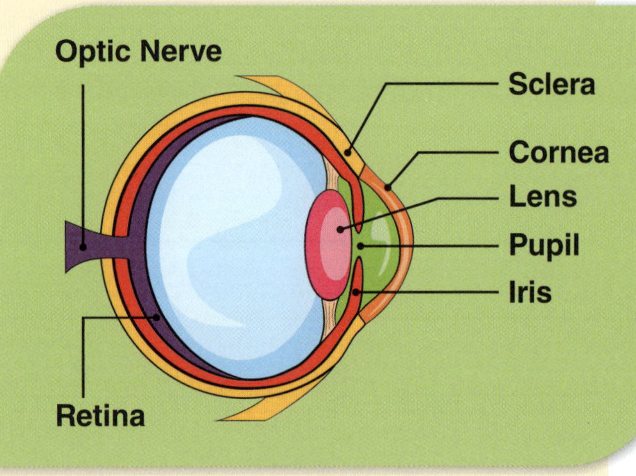

Answer the questions.

1. What shape are our eyes? ______________

2. Which part of the eye controls its movement? ______________

Eyesight

TARGETING SCIENCE YEAR 5 © PASCAL PRESS ISBN: 9781925726541

Light Enters the Eye

Concepts:

Light reflects off objects and travels in a straight line into the eye.

Light enters the eye through the cornea to the pupil.

Define It!

cornea: the clear layer forming the front of the eye

reflect: to throw back heat, light, or sound without absorbing it

refract: to change the direction of light

To see the world around you, your eyes need light. Eyes send your brain information about an object's shape, colour, and movement. They do this by taking in light that **reflects** off that object. Light reflected off an object travels in a straight line into the eye.

Light enters the eye through the **cornea**. The cornea is a clear coating that covers the iris and pupil at the front of the eye. It helps to **refract**, or change the direction of the light. It also helps the eye to focus. Light travels through the cornea to the pupil. When it is very bright and there is a lot of light, the pupil is small. When it is dark, the pupil grows larger in order to let more light into the eye.

Write *true* or *false*.

1. Light travels in a straight line into the eye. ____________
2. The pupil grows larger when it is bright out. ____________
3. The cornea helps to refract light. ____________

How We See

Concepts:

The lens refracts light to the retina.

The retina changes images into nerve signals that travel through the optic nerve to the brain.

Define It!

lens: the part of the eye that focuses and refracts light

optic nerve: nerves that send signals from the eye to the brain

project: to cause light to appear on a surface

retina: a layer at the back of the eye that changes images into nerve signals

After light passes through your pupil, it travels to the **lens**. The lens helps to focus the light to a part on the back of the eye called the **retina**. Because the lens refracts light, the image **projected** to the retina is upside down! The retina's job is to change the projected image into nerve signals. These signals are sent through the **optic nerve** to the brain. The brain is then able to make sense of what you are seeing.

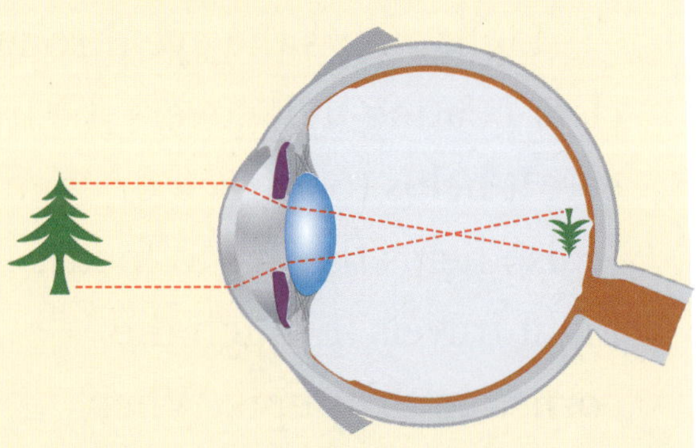

Say you are looking at a tree. In an instant, you are able to tell that it is a tree because light reflected off the tree and travelled in a straight line into your pupil. The lens refracted the light and projected the image to your retina. The retina changed the image into nerve signals, which travelled along the optic nerve to your brain. Then your brain told you, "It's a tree!"

Complete the sentences.

1. The ____________________ sends signals to the brain.

2. The lens projects an image onto the ____________________.

Eyesight

TARGETING SCIENCE YEAR 5 © PASCAL PRESS ISBN: 9781925726541

The Eyes Have It

Skills:

Interpret and identify information in graphic representations.

Look at the numbered diagram of the eye. Next to each matching number and description below, label the *cornea, iris, lens, optic nerve, pupil, retina,* and *sclera*.

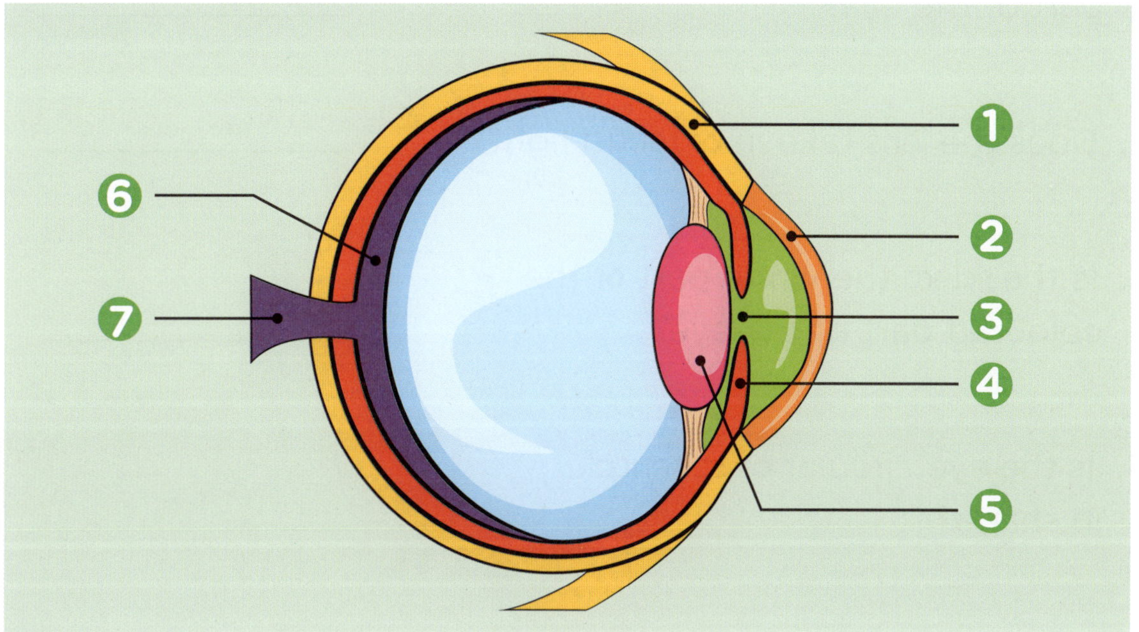

1. moves the eye ______________
2. refracts light and helps the eye focus ______________
3. takes in light ______________
4. controls the amount of light that enters the eye ______________
5. refracts light and focuses it to the back of the eye ______________
6. transforms images into nerve signals ______________
7. sends nerve signals to the brain ______________

Skill:

Apply content vocabulary.

Either/Or Questions

Write each answer.

1. Is the pupil **or** the iris the coloured part of the eye? ____________________

2. Does the lens take in **or** refract light? ____________________

3. Is the sclera the white part **or** the coloured part of the eye? ____________________

4. Is the eye circular **or** spherical in shape? ____________________

5. Is an image projected **or** reflected onto the retina? ____________________

6. Does the cornea **or** the optic nerve send signals to the brain? ____________________

TARGETING SCIENCE YEAR 5 © PASCAL PRESS ISBN: 9781925726541

Blind Spot

Skills:

Conduct experiments and draw conclusions about the results.

Have you ever heard adults complain about a "blind spot" while driving? Perhaps they didn't see an oncoming car while they were changing lanes. We all have blind spots. Complete the activities to see (or rather, *not* see!) your blind spot in action.

What You Need

- 8 cm x 13 cm index card (or other stiff paper)
- pen or marker
- ruler

Directions

1. Mark a dot and an **X** on the card as shown.

2. Hold the card at eye level and about an arm's length away. Make sure that the **X** is on the right side.
3. Close your right eye and look directly at the **X** with your left eye. Notice that you can also see the dot. Focus on the **X** but be aware of the dot as you slowly bring the card closer to your face.
4. Now close your left eye and look directly at the dot with your right eye. As you slowly bring the card closer to your face, notice what happens to the **X**.
5. Repeat Steps 2–4, but this time hold the card so that it is at an angle.

Eyesight

6. Using the ruler, draw a straight line through the dot and the **X** to both edges of the card.

7. Repeat Step 3. Notice what happens to the line.

What Did You Discover?

1. What happened to the dot in Step 3?

___

2. What happened to the **X** in Step 4?

___

3. Were the results the same in Step 5? If not, explain your answer.

___

4. What happened to the dot and the line in Step 7?

___

What Happened?

The optic nerve passes through one spot on the retina. In this spot, the retina cannot receive light. When you hold the card so that the light from the dot or **X** falls on this spot, you cannot see the dot or the **X**. However, the line does not disappear, because your brain automatically "fills in" the blind spot. It does this by using information from the line that continued to the edges of the card.

TARGETING SCIENCE YEAR 5 © PASCAL PRESS ISBN: 9781925726541

Skill:

Answer text-based questions.

Choose an object in your room and focus on it. Now answer the questions to describe how you are able to see this object.

1. What is the object?

____

2. How does light enter your eye from the object?

____

____

3. Through which two parts of the eye does light enter?

____

4. Which part of the eye refracts the light onto the retina?

____

5. How does the image of your object appear on your retina?

____

6. What is the retina's job?

____

7. Which part of the eye sends nerve signals to the brain?

____

8. What happens when nerve signals reach your brain?

____

What is matter?

Matter is the 'stuff' which everything is made of. It is anything that takes up space, or can be weighed. Matter is made up of atoms and molecules.
Your teeth, the clothes you are wearing, the seat you are sitting on and the air you are breathing ... it is all matter.
Matter can take one of four states: solid, liquid, gas or plasma. In this unit, we will be focusing on solids, liquids and gases.
Matter can be classified in these groups by the properties it demonstrates. If you can learn to understand these properties and what they mean, it will help you to be able to classify matter into the correct group.

Solids	Liquids	Gases
• Solids hold their shape (they stay the same if you walk away and come back – even over a very long period of time). This is because the molecules which they are made up of are tightly packed together. • Solids do not flow, even if left for a long period of time. • Solids form a mound (or pile) when you pour them, like sugar on a plate.	• Liquids take the shape of the container they are in. If the bowl is round, the liquid will take that shape. This is because the molecules in liquids are not as tightly packed together as they are in solids, so they can move around more within the liquid. • Liquids flow, but some of them flow VERY slowly. • Liquids stay level as they fill a container.	• Gases take the shape of the container they are in, but if you take the lid off, they will also flow out of the container. This is because the molecules in gases can flow freely, and will spread all through the space they are in.

DENSITY OF MATTER

GAS

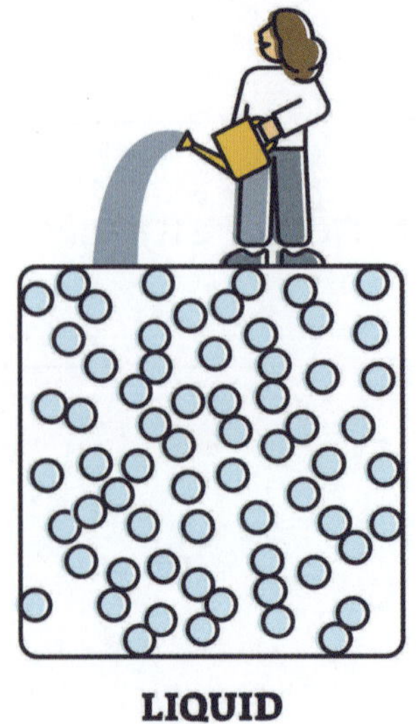

TARGETING SCIENCE YEAR 5 © PASCAL PRESS ISBN: 9781925726541

https://clickv.ie/w/Dwgx

Use this QR code to access a video on this topic.

Define It!

matter: an object or substance that has mass and takes up space

particle: a very small piece of something

physical property: a trait, such as colour or shape, that describes a substance

state: a way of living or existing

substance: a particular kind of matter with the same properties

Nearly everything in the universe is made of matter. This paper is made of **matter**. You are made of matter. Even the air we breathe is made of matter. But what is matter? Matter is any object or **substance** that is composed of **particles** and takes up space.

Matter can exist in three **states**: solid, liquid, and gas. Each state has its own set of **physical properties**. This means that a solid looks, feels, and acts differently than a liquid or a gas. A solid keeps its own shape, and its particles do not move around and are tightly packed together. A liquid does not keep its own shape; it takes the shape of its container. The particles of a liquid move around and are farther apart from each other than in a solid. Finally, a gas is usually invisible and has no shape. Its particles move around freely and far apart from one another.

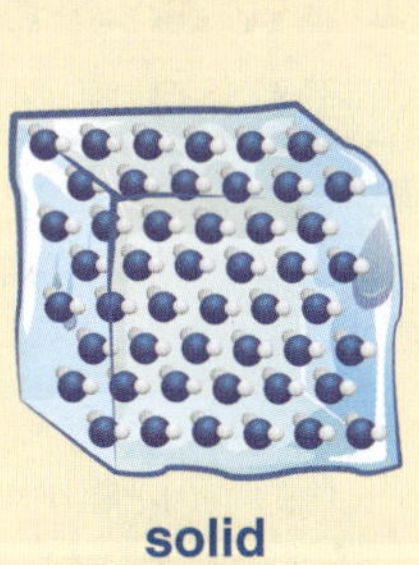

solid

gas

liquid

Answer the question.

A can of soft drink contains matter in all three states. Which part is a solid? Which part is a liquid? Which part is a gas?

____

____

The Building Blocks of Matter

An atom is the smallest part of matter. It is the basic 'building block' for all of the matter in the universe. Atoms are very, very small. You could fit about 10 million hydrogen atoms just into the head of a pin!

An element is a substance that is made up of only one kind of atom. Some examples of elements are helium, carbon and oxygen. There are 118 different elements that have been recognised on Earth, and all matter is made of them in some way.

Periodic Table of the Elements

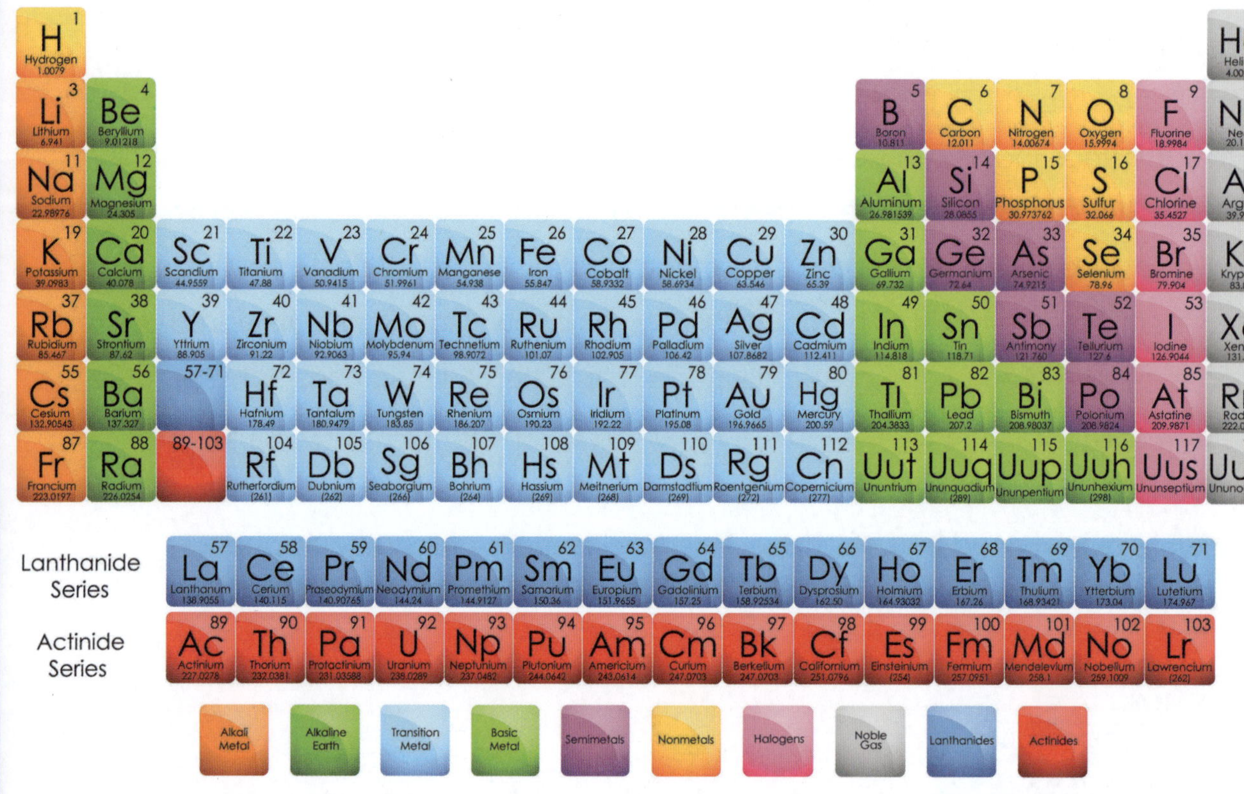

A molecule is made up of either one kind or more than one kind of atom. A molecule of water has two atoms of hydrogen, and one atom of oxygen. The gas oxygen is made up of two oxygen atoms.

Solid, liquid or gas?

Using the guide on page 83, can you classify these substances into solids, liquids and gases? Write what they are in each box under the picture. Use S for solids, L for liquids and G for gases.

Gases

Try these simple experiments to test out the properties of gases.

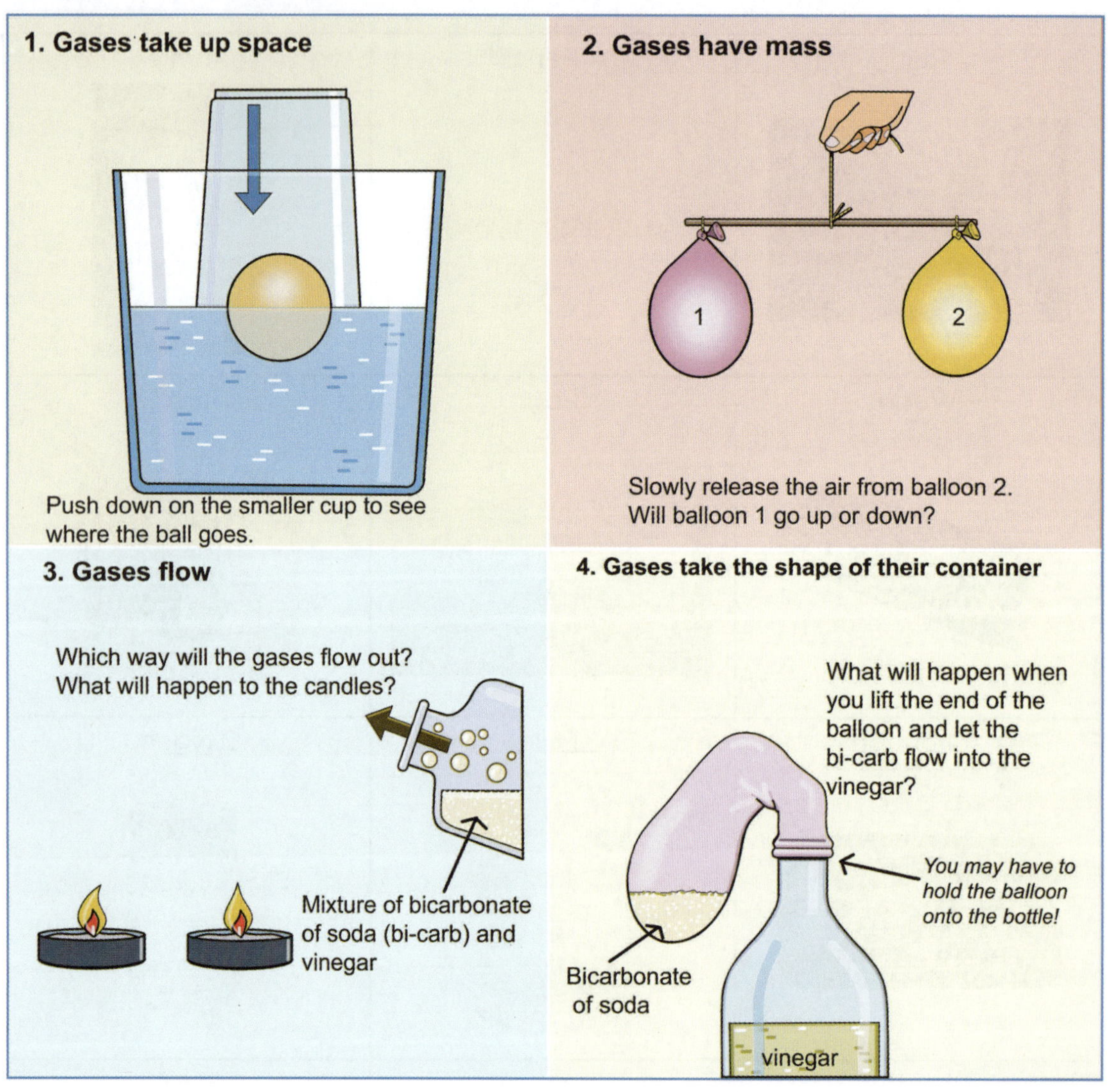

Circle *true* or *false*.

1. Gases stay in one place: **true** **false**
2. Gases take up space in a container: **true** **false**
3. Gases can be compressed: **true** **false**
4. Gases do not flow: **true** **false**
5. Gases weigh something: **true** **false**

Matter

TARGETING SCIENCE YEAR 5 © PASCAL PRESS ISBN: 9781925726541

Compressing Solids Liquids and Gases

Compression is a force that squeezes something together. When something is compressed, it takes up less room.

Gas particles have a lot more space between them than the particles in a solid. This space, between particles, is called **interparticular or intermolecular space.**

In the first syringe, there is a solid. In the second syringe there is a liquid and in the third syringe there is a gas.

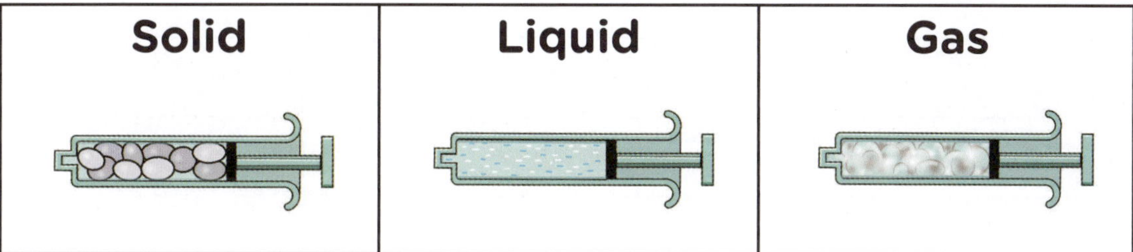

If you were able to see these under a powerful microscope, the molecules would look like this (but they would be much smaller).

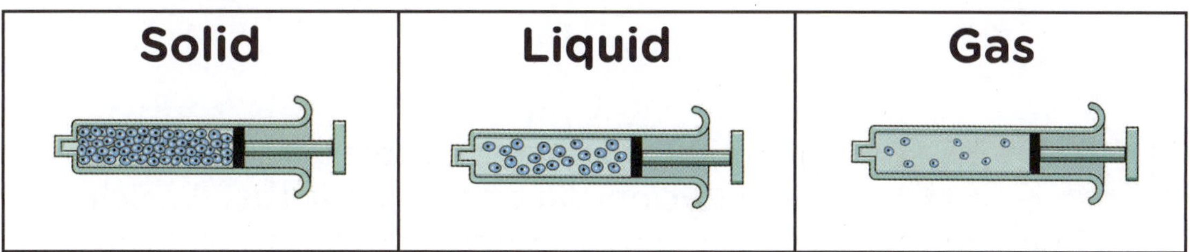

What will happen as you try to compress each syringe? Why?

____

____

____

____

Name the Properties

Choose two objects in your house to observe. Look at them both closely and write down as many observations about the objects as you can. Then touch the objects and write down what you feel. Make a star next to the observations that are *physical properties*.

Object 1	Object 2

Challenge! Look up the physical properties of your objects in an encyclopedia or on the Internet. Which physical properties did you correctly identify? Which ones did you miss? What kind of physical properties do your objects have that are invisible but still able to be measured?

 TARGETING SCIENCE YEAR 5 © PASCAL PRESS ISBN: 9781925726541

Define It!

boiling point: the temperature at which a liquid changes into a gas

freezing point: the temperature at which a liquid changes into a solid

physical change: a change in which the substance remains the same substance

Matter does not always stay the same—it can change. You can slice a piece of cheese in half or scratch a mirror. The cheese and the mirror have gone through a **physical change**, but they did not become different substances. A scratched mirror is still a mirror, and a smaller piece of cheese is still cheese.

Matter can also change its state from solid to liquid to gas. Change in temperature is the most common cause for matter to change states. The temperature at which a substance changes from a liquid to a solid is called its **freezing point**. For example, the freezing point of water is 0°C. At this temperature, rain turns to snow and liquid water turns to ice. The temperature at which a substance changes from a liquid to a gas is called its **boiling point**. The boiling point of water is 100°C. At this temperature, liquid water evaporates into a gas called water vapour. Freezing point and boiling point are both physical properties. Different substances have different freezing and boiling points.

Concepts:

Matter can change state from solid to liquid to gas.

Physical changes do not create different substances.

Room temperature is approximately 20°C.

	Freezing Point	Boiling Point
Water	0°C	100°C
Nitrogen	-210°C	-196°C
Mercury	-39°C	357°C
Gold	1,064°C	2,856°C

Use the information above to answer the questions.

1. Which substance has the highest boiling point? ________________

2. Which substance has the lowest freezing point? ________________

3. Which two substances are liquids at room temperature?

________________ ________________

TARGETING SCIENCE YEAR 5 © PASCAL PRESS ISBN: 9781925726541

Other Physical Properties

Concepts:

Physical properties include colour, hardness, texture, ability to conduct heat, and response to magnetic forces.

Define It!

conduct: to move or transfer energy

magnetic: capable of being attracted by a magnet

observe: to see and notice something

Boiling point and freezing point are not the only physical properties of a substance. Physical properties may include colour, hardness, texture, or many other characteristics of matter that can be observed or measured. If you were to look at a rock, what would you see? You might describe its colour and shape, or whether it's shiny or dull. If you picked it up, you might feel whether it's smooth or rough. These all describe physical properties of the rock.

Some physical properties are actually invisible, but you can still **observe** or measure their effects. For example, an object's ability to **conduct** heat or electricity is a physical property. Copper is a good conductor of electricity and heat, which is why copper is used in wires and pots and pans. A substance's response to **magnetic** forces is also a physical property. For example, certain metals, such as iron, are attracted by magnets. Other metals, such as copper and aluminium, are not attracted by magnets.

Answer the questions.

1. Name three physical properties that you can observe or measure.

____

____

2. Name two physical properties that are invisible but still can be observed.

____

____

TARGETING SCIENCE YEAR 5 © PASCAL PRESS ISBN: 9781925726541

Ruby Red

Skill: Interpret information from graphic images.

Look at the two pictures of a ruby below. One picture shows the stone raw, meaning this is the way it looks when it comes out of the ground. The other picture shows the ruby cut and polished into a gem, which is how it's usually used in jewellery. Use the pictures to answer the questions.

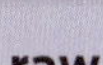
raw

polished

1. Describe the physical properties of the raw ruby stone.

2. Describe the physical properties of the cut and polished ruby gem.

3. Did the ruby undergo a physical change? Explain your answer.

Vocabulary Practise

Skill:

Apply content vocabulary.

Select from the list of vocabulary words to complete the crossword puzzle.

boiling point	conduct	substance	magnetic
freezing point	observe	particle	physical change
physical property	state	matter	

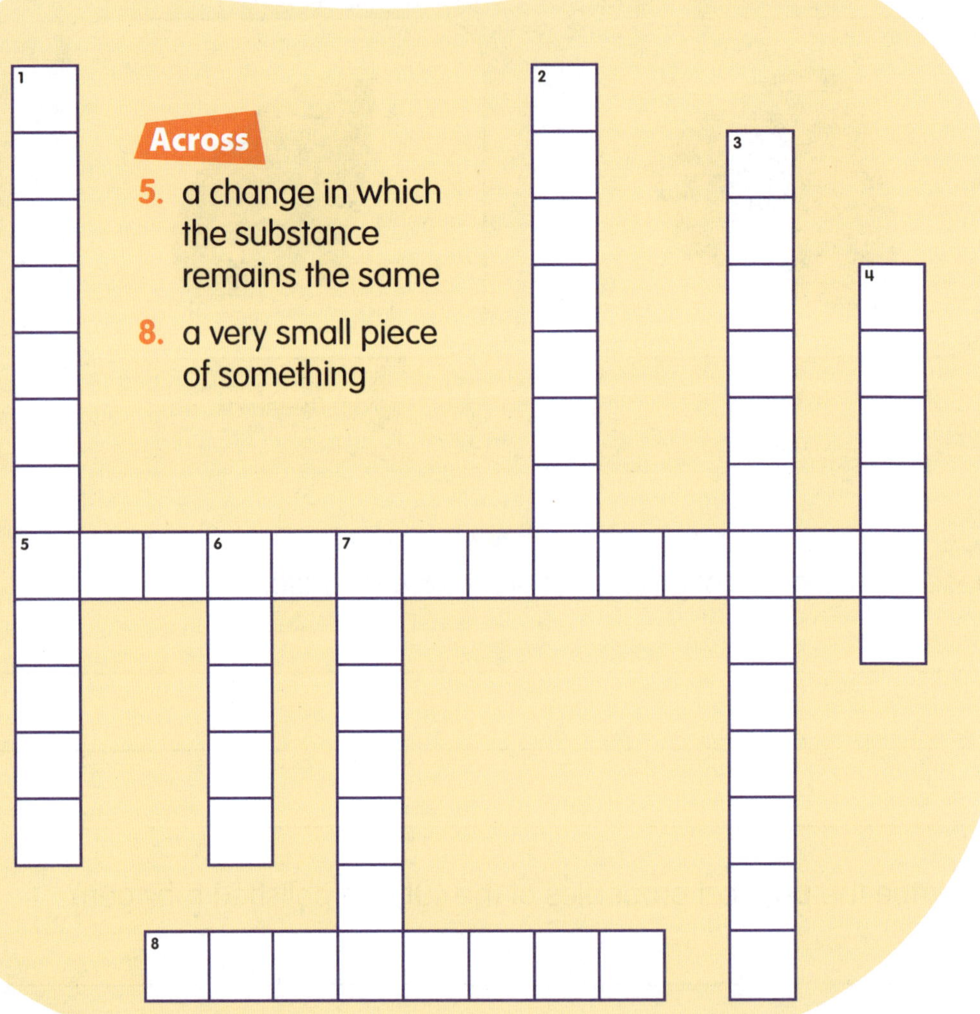

Across

5. a change in which the substance remains the same
8. a very small piece of something

Down

1. the temperature at which a liquid changes into a gas
2. capable of being attracted by a magnet
3. the temperature at which a liquid changes into a solid
4. an object or substance that has mass and takes up space
6. a way of existing
7. to transfer energy

TARGETING SCIENCE YEAR 5 © PASCAL PRESS ISBN: 9781925726541

A Matter of Change

In this activity, you will make changes to several different objects. Observe and describe what the objects looked like before you made the change and what they looked like after. Then answer the questions to test your understanding of physical properties.

Skills:

Conduct experiments, record data, and analyse results.

What You Need

- sheets of paper
- pencil
- pencil sharpener
- marshmallow
- chocolate bar
- hot plate or micro-wave (to be used with a parent or guardian's help)

Directions

1. Record all of your observations in the "Before change(s)" column of the chart on page 94 before you make changes to the objects.
2. Tear the first sheet of paper in half. Crumple the second sheet of paper into a ball. Record your observations.
3. Sharpen the pencil into a fine point and remove the pencil shavings from the pencil sharpener. Record your observations.
4. Have an adult help you heat the marshmallow in the microwave for 10 to 15 seconds. Record your observations.
5. Then have the adult help you heat the chocolate bar in the microwave for 30 to 45 seconds. Record your observations.

A Matter of Change

	Before change(s)	After change(s)
Paper (torn)		
Paper (crumpled)		
Pencil		
Marshmallow		
Chocolate		

1. How did the papers change? Was anything new created?

2. How did the pencil change? Was anything new created? Is it the same substance?

3. How did applying heat to the marshmallow and the chocolate change them? Was a new substance created?

4. What can you say about the types of changes that took place in all of these objects? Explain your answer.

TARGETING SCIENCE YEAR 5 © PASCAL PRESS ISBN: 9781925726541

First Nations people demonstrated that they had a solid understanding of how matter can could change states, long before the colonisation of Australia.

Procedures including the extraction of oils (solid-liquid), medicine making (liquid-gas) and cooking practises (liquid-gas) were used. These are just a few examples.

On the western Cape York Peninsula, the Anguthimri and the Awngthim Peoples heated stingray fat in a large bailer shell.

The liquid oil has long been used as a treatment in the manufacture and repair of wooden implements such as the spear-thrower.

The Wiradjuri People of central New South Wales treated colds by building steam pits that were heated by fires, lined with Eucalyptus leaves and overlaid with possum-rugs. The Eucalyptus oil was turned into steam through this process and inhaled as a decongestant treatment for coughs and colds.

Changes to the States of Matter from a First Nations Perspective

The Barindji Peoples of south central New South Wales cooked large animals in a ground oven into which hot stones were placed. The animal was then put into the oven, covered with grass, more hot stones were placed on top, and it was completely enclosed with soil. At various cooking intervals a hole was made in the oven and water was poured in to steam the food. The addition of water to the oven demonstrates the knowledge that the steaming process requires water, as a liquid to generate steam for cooking.

1. How do you think young First Nations people learned to cook using these methods?

___

___

___

2. Have you ever used Eucalyptus when you have a cold? Was it used as a solid, liquids or gas? Explain your answer.

___

___

___

TARGETING SCIENCE YEAR 5 © PASCAL PRESS ISBN: 9781925726541

Page 2

1 structural; 2 structural; 3 behavioural

Page 3

1 By camouflage, that is, its colour blends in with its surroundings.

2 The dark-coloured mice.

Page 4

1 true, 2 false, 3 true

Page 5

1 spread out like a net to quickly drink up the rain.

2 can swell and store many litres of water.

Page 6

Structural adaptations: frog's skin is able to hold water; frog has strong back legs to help it burrow underground; frog has strong front legs for catching prey.

Behavioural adaptations: frog digs under the ground after rain; frog only emerges when it has rained; male water holding frogs make loud, long, slow 'maaaw-w-w' sounds to attract females.

Page 7

Adaptations to keep cool in the heat (they sweat and pant and lick their forearms), pouches to keep joeys cool, ability to hop to move through their environment (enlarged hind feet and strong tail), search for food at dusk and dawn to avoid heat of the day.

Page 8

1 It was limp.

2 It was straight and tall.

3 They both drink up and store water. A cactus has spines but celery does not.

Page 10

Pandas: live in China, eat bamboo; Both: part of the bear family; Polar bears: live in the Arctic circle, eat seals

Page 11

1 Flat teeth help pandas grind bamboo and a wrist bone helps pandas hold bamboo.

2 Small bumps on polar bears' paws keep them from slipping on the ice. Thick claws help grip the ice and hold prey. Webbed toes help polar bears swim.

Page 12

Pandas are in danger because their habitat is being destroyed, which means there is less bamboo for them to eat. Polar bears are in danger because the ice caps are melting, which means they have less space to hunt seals.

Page 13

1 They look like dead leaves; they hang motionless from branches; the eggs look like seeds; the nymphs look like ants.

Page 14

1 orange-bellied parrot; 2 smaller; 3 humans and koalas; 4 bigger; 5 thylacine

Page 15

1 blubber; 2 habitat, extinction; 3 adaptations, specialised; 4 diet. Puzzle word: tundra

Page 17

1 The water felt very cold without the glove. I could not feel the coldness of the water when I the glove on my hand.

2 No, I could not feel the heat of the washcloth.

3 The polar bear would get hot very quickly.

Page 18

Answers will vary, but should show an understanding of the concept of specialised adaptations plus the specific adaptations that polar bears and pandas make to their environment.

Page 21

1 true; 2 true; 3 false

Page 22

1 herd; 2 deep, rumbling; 3 danger

Page 23

1 They go to warmer places that have more food.

2 They honk to let each other know where they are.

Page 24

1 Closer to the coast, where it is warmer.

2 They gather together in large clusters and hang from branches until it gets warmer.

3 December, January and February.

Page 25

1 pride; 2 prey; 3 habitat; 4 trumpeting; 5 charge; 6 rumbling; 7 migrate; 8 flock; 9 carnivore

Page 28

1 Their parents are not around to keep them safe.

2 One and a half years.

Page 29

1 white; 2 orange

Page 30

1 tadpole; 2 cygnet; 3 hind; 4 trait; 5 cub; riddle: baby chick

Answer Key

Page 32

1 smooth rock on the beach; 2 sand dunes

Page 33

1 1.6 km; 2 5-6 million years ago

Page 34

1 storms, earthquakes, landslides; 2 the amount and composition of soil

Page 35

Gravity, rainwater, wind, ice.

Page 36

Answers will vary but should show an understanding of how erosion and weathering brought on the change from a mountainous area to a deep, wide canyon with a river running through it.

Page 37

A examples: orange with red and white swirls, smooth, oval.

B examples: brownish-orange, rough, misshapen

Rock A looks more weathered because it is smooth, and smoothness is due to weathering.

Page 38

1 sloping; 2 weathering; 3 landform; 4 channels; 5 expand; 6 spectacular; 7 sediment; 8 erosion

Page 39

2 The best guess is the soil structure, because soil sticks together better.

Page 40

1 likely the dry sand fell down first and the mud withstood the wind

2 likely the mud structure

Page 41

1, 3, 4, 2

Page 42

1 Erosion from the Colorado River carried away earth.

2 Frost wedging pushed rocks apart.

3 Gravity caused parts of the canyon wall to collapse.

4 Wind blew sand that chipped away at the canyon walls.

Page 43

1 Floodwaters carry away rocks and sand that then block parts of the river.

2 It stopped natural flooding, which meant that rocks and sand that blocked parts of the river weren't carried away.

3 Without natural flooding, sandbars wouldn't form. That means animals and plants wouldn't have them as a habitat.

Page 44

1 It is raining, which means that rainwater could be seeping into the cracks of the canyon wall.

2 It is colder, which means that the rainwater could have frozen and expanded, pushing the rocks apart.

3 Gravity is pulling down large chunks of the canyon wall.

Page 47

1 Answers will vary, but should show an understanding of the arguments for and against building near sand dunes.

Page 48

1 weathering; 2 landform; 3 habitat; 4 sandbar; 5 channels; 6 geosphere; 7 erosion

Page 50

1 Tiny rocks and sand. It was dry and coarse.

2 The dirt sunk down where the water was poured.

3 Some water absorbed into the soil and some dripped into the pan below.

4 If the soil was already wet, less water would absorb and more soil would erode.

5 More dirt would flow out with the water.

Page 52

1 They can flow over their banks.

2 They may be destroyed.

Page 53

1 false; 2 false; 3 true

Page 54

1 Colorado River; 2 reservoir; 3 electricity

Page 55

1 arch dam; 2 buttress dam; 3 gravity dam; 4 embankment dam

Page 57

You can make the yarn unwind and lower the pencil by holding the waterwheel so it turns in the opposite direction.

Page 58

1 Floods destroy crops and property, lives may be lost, it costs billions of dollars to rebuild.

2 The flooding Nile brought silt that was rich in nutrients, the new soil was good for growing crops.

TARGETING SCIENCE YEAR 5 © PASCAL PRESS ISBN: 9781925726541

Page 59

natural: firefly, stars, bushfire, lightning, sun

artificial: lamp, torch, fireworks, traffic light, light house, light bulb, car headlights

Page 60

1 visible light; 2 300,000 km per second

Page 61

1 false; 2 true; 3 false

Page 62

hand - dark; magnifying glass - very faint; tissue - lighter; wheelbarrow - dark; flyscreen - lighter; glass - very faint

Page 63

1 reflect light; 2 refracted the light

Page 64

1 transparent; 2 translucent; 3 water in pool, chair legs, drinking glasses, sunglasses, table; 4 ground, woman, chairs, towel, umbrella, hat, bathing suit; 5 water in pool, drinks in glasses

Page 68

The pencil appeared bent when looked at from the side - ie, through the water. This is due to refraction, and the pencil was not really bent or broken. It did not appear bent when it was standing straight and looked at directly from above, as the light rays did not pass through the water first.

Page 70

1 transparent; 2 absorb; 3 laser; 4 opaque; 5 translucent; 6 reflect; 7 (down) ray; 7 (across) refract

Page 72

4 answers will vary

6 The water in the jars got warmer.

7 The water in the jar covered with black paper got warmer.

8 The jars absorbed the energy in sunlight.

9 black paper

Page 73

Answers will vary but should refer to somethings that are opaque, translucent, and transparent. They should also mention electromagnetic waves, reflection or refraction.

Page 74

1 spherical; 2 sclera

Page 75

1 true; 2 false; 3 true

Page 76

1 optic nerve; 2 retina

Page 77

1 sclera; 2 cornea; 3 pupil; 4 iris; 5 lens; 6 retina; 7 optic nerve

Page 78

1 iris; 2 refract; 3 white; 4 spherical; 5 projected; 6 optic nerve

Page 80

1 answers will vary; 2 light reflects off object and enters eye in a straight line; 3 cornea and pupil; 4 lens; 5 upside down; 6 to change the image into nerve signals; 7 optic nerve; 8 your brain tells you what you are seeing

Page 83

The can is a solid, the soft drink is a liquid, the bubbles are a gas.

Page 85

L	S	G
S	G	L
S	L	G
S	L	S
S	L	G

Page 86

1 false; 2 true; 3 true; 4 false; 5 true

Page 87

1 The solid compressed the least, then the liquid. The gas compressed the most. This is because the molecules are closest together in a solid, and furthest apart in a gas.

Page 89

1 gold; 2 nitrogen; 3 water and mercury

Page 90

1 colour, hardness, texture; 2 ability to conduct heat or electricity, response to magnetic forces

Page 91

1 scratched, dull, lumpy, pinkish, etc; 2 shiny, sparkly, magenta, circular with many cuts, etc;

3 yes, it did not become a different substance, it is still a ruby

Page 92

1 boiling point; 2 magnetic; 3 freezing point; 4 matter; 5 physical change; 6 state; 7 conduct; 8 particle

Answer Key

Page 94

1 The papers were smoother and larger before I tore and crumpled them. Nothing new was created.

2 The pencil became sharp and shorter. Shavings were created, but they are the same substance as the pencil.

3 The marshmallow expanded and the chocolate melted, but they were still the same substance.

4 The matter in the objects changed shape, size or texture, but no new substances were created. This means that the changes were all physical changes.

Page 96

1 By watching their parents and elders.

2 Eucalyptus lollies - solid, eucalyptus essential oil - liquid, eucalyptus inhaler - gas.

TARGETING SCIENCE YEAR 5 © PASCAL PRESS ISBN: 9781925726541